Springer Undergraduate Mathematics Series

Advisory Board

Other books in this series

John M. Howie

Real Analysis

With 35 Figures

 Springer

John M. Howie, CBE, MA, D.Phil, DSc, Hon D.Univ, FRSE
School of Mathematics and Statistics, Mathematical Institute,
University of St Andrews, North Haugh, St Andrews, Fife, KY16 9SS, Scotland

Cover illustration elements reproduced by kind permission of:

Aptech Systems, Inc., Publishers of the GAUSS Mathematical and Statistical System, 23804 S.E. Kent-Kangley Road, Maple Valley, WA 98038, USA. Tel: (206) 432 - 7855 Fax (206) 432 - 7832 email: info@aptech.com URL: www.aptech.com

American Statistical Association: Chance Vol 8 No 1, 1995 article by KS and KW Heiner 'Tree Rings of the Northern Shawangunks' page 32 fig 2

Springer-Verlag: Mathematica in Education and Research Vol 4 Issue 3 1995 article by Roman E Maeder, Beatrice Amrhein and Oliver Gloor 'Illustrated Mathematics: Visualization of Mathematical Objects' page 9 fig 11, originally published as a CD ROM 'Illustrated Mathematics' by TELOS: ISBN 0-387-14222-3, German edition by Birkhauser: ISBN 3-7643-5100-4.

Mathematica in Education and Research Vol 4 Issue 3 1995 article by Richard J Gaylord and Kazume Nishidate 'Traffic Engineering with Cellular Automata' page 35 fig 2. Mathematica in Education and Research Vol 5 Issue 2 1996 article by Michael Trott 'The Implicitization of a Trefoil Knot' page 14.

Mathematica in Education and Research Vol 5 Issue 2 1996 article by Lee de Cola 'Coins, Trees, Bars and Bells: Simulation of the Binomial Process' page 19 fig 3. Mathematica in Education and Research Vol 5 Issue 2 1996 article by Richard Gaylord and Kazume Nishidate 'Contagious Spreading' page 33 fig 1. Mathematica in Education and Research Vol 5 Issue 2 1996 article by Joe Buhler and Stan Wagon 'Secrets of the Madelung Constant' page 50 fig 1.

British Library Cataloguing in Publication Data
Howie, John Mackintosh
  Real analysis. - (Springer undergraduate mathematics
series)
  1. Mathematical Analysis
  I. Title
  515
ISBN 1852333146

Library of Congress Cataloging-in-Publication Data
Howie, John M. (John Mackintosh)
  Real analysis / John M. Howie
    p. cm. -- (Springer undergraduate mathematics series, ISSN 1615-2085)
  Includes bibliographical references and index.
  ISBN 1-85233-314-6 (alk. paper)
  1. Mathematical analysis. I. Title. II. Series.
QA300.H694 2001
515—dc21   00-069839

Springer Undergraduate Mathematics Series ISSN 1615-2085
ISBN 1-85233-314-6 Springer-Verlag London Berlin Heidelberg
Springer Science+Business Media
springeronline.com

Typesetting: Camera ready by the author
Printed and bound at the Athenæum Press Ltd., Gateshead, Tyne & Wear
12/3830-5432 Printed on acid-free paper SPIN 11338116

To my grandchildren
Catriona, Sarah, Karen and Fiona,
who may some day want to read this book

Preface

From the point of view of strict logic, a rigorous course on real analysis should precede a course on calculus. Strict logic, is, however, overruled by both history and practicality. Historically, calculus, with its origins in the 17th century, came first, and made rapid progress on the basis of informal intuition. Not until well through the 19th century was it possible to claim that the edifice was constructed on sound logical foundations. As for practicality, every university teacher knows that students are not ready for even a semi-rigorous course on analysis until they have acquired the intuitions and the sheer technical skills that come from a traditional calculus course.

Real analysis, I have always thought, is the *pons asinorum*[1] of modern mathematics. This shows, I suppose, how much progress we have made in two thousand years, for it is a great deal more sophisticated than the Theorem of Pythagoras, which once received that title. All who have taught the subject know how patient one has to be, for the ideas take root gradually, even in students of good ability. This is not too surprising, since it took more than two centuries for calculus to evolve into what we now call analysis, and even a gifted student, guided by an expert teacher, cannot be expected to grasp all of the issues immediately.

I have not set out to do anything very original, since in a field as well-established as real analysis originality too easily becomes eccentricity. Although it is important to demonstrate the limitations of a visual, intuitive approach by means of some "strange" examples of functions, too much emphasis on "pathology" gives altogether the wrong impression of what analysis is about. I hope that I have avoided that error.

It is, of course, possible to handle many of the fundamental ideas of real analysis within the more general framework of metric spaces. Assuredly this has advantages, but I have preferred to take the more concrete approach, believing

---

[1] The bridge of asses; that is, the bridge between elementary mathematics, which asses can understand, into higher regions of thought.

as I do that to add abstraction to the difficulties facing the student is to make the subject unnecessarily daunting. By the same token, though many of the notions apply to both real and complex numbers, it seemed that there was little to gained by drawing attention to this aspect.

My experience suggests that the central notion of a limit is more easily approached in the context of sequences, and I take this as a starting point. Sequences, and the closely related topic of series, occupy Chapter 2. Chapter 3 (Continuity), Chapter 4 (Differentiation) and Chapter 5 (The Riemann Integral) form the core of the book. Chapter 6 introduces the logarithmic and exponential functions, and the crucial concept of uniform convergence is introduced in Chapter 7 (Sequences and series of functions).

Any author setting out to write a book on real analysis has to decide where to introduce the circular functions sin and cos. Logically one ought to delay their introduction until it can be done "properly", and this is what I would do if I ever wished (perish the thought!) to write a rigorous *Cours d'Analyse* in the grand French manner. There must, however, be few students setting out on a first course in analysis who are unfamiliar with sines and cosines, and I am certainly not the first author to adopt the practical policy of regarding these functions as provisionally "known", pending proper definitions. The proper definitions are given in Chapter 8.

It is all too easy to present mathematics as a set of truths inscribed on tablets of stone, complete and perfect. By various footnotes and index entries I have sought to emphasise that the subject was created by real people. Much interesting information on these people can be found in the St Andrews History of Mathematics archive (http://www-history.mcs.st-and.ac.uk/history/).

The book contains many worked examples, as well as 190 exercises, for which brief solutions are provided at the end of the book.

I retired in 1997 after 27 years in the University of St Andrews. It is a pleasure to record thanks to the University, and in particular to the School of Mathematics and Statistics, for their generosity in continuing to give me access to a desk and to the various facilities of the Mathematical Institute.

I am grateful to my colleague John O'Connor for his help in creating the diagrams. Warmest thanks are due also to my colleagues Kenneth Falconer and Lars Olsen, whose comments on the manuscript have, I hope, eliminated serious errors. The responsibility for any imperfections that remain is mine alone.

John M. Howie
University of St Andrews
November, 2000

Contents

1
Introductory Ideas

1.1 Foreword for the Student: Is Analysis Necessary?

In writing this book my assumption has been that you have encountered the fundamental ideas of analysis (function, limit, continuity, differentiation, integration) in a standard course on calculus. For many purposes there is no harm at all in an informal approach, in which a continuous function is one whose graph has no jumps and a differentiable function is one whose graph has no sharp corners, and in which it is "obvious" that (say) the sum of two or more continuous functions is continuous. On the other hand, it is not obvious from the graph that the function f defined by

$$f(x) = \begin{cases} x\sin(1/x) & \text{if } x \neq 0 \\ 0 & \text{if } x = 0 \end{cases}$$

is continuous but not differentiable at $x = 0$, since the function takes the value 0 infinitely often in any interval containing 0, and so it is not really possible to draw the graph properly. More significantly, since the sine function is continuous it might seem obvious that the function S defined by

$$S(x) = \frac{\sin x}{1} - \frac{\sin 2x}{2} + \frac{\sin 3x}{3} - \cdots$$

is continuous, but in fact its graph is

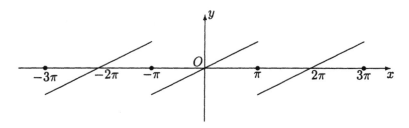

(The dots on the x-axis indicate that the function takes the value 0 at these points.)

Interestingly, paradoxes of this kind came to light in the context of what we would now call applied mathematics, when Fourier[1] was studying heat conduction. At the time, the function S with the above graph would not have been thought of as a function at all, and yet it was defined as an infinite series of perfectly respectable functions. It was a real crisis, and forced mathematicians into some serious thinking about functions, limits, continuity and convergence. This happened throughout the nineteenth century, and by the dawn of the twentieth century it could be fairly claimed that calculus was at last built on secure logical foundations.

This "new mathematics", developed mostly in Germany and France, took a little time to filter down into British undergraduate syllabuses, but the publication in 1908 of Hardy's[2] "A Course of Pure Mathematics" proved a watershed, and a course on "rigorous" analysis is now regarded as an essential component of an honours undergraduate course. Assuredly most of the dramatic developments of mathematics in the twentieth century could not have taken place without the secure foundations provided by analysis. It takes some effort to master it, but the reward is the excitement that comes from understanding and appreciating one of the supreme achievements of the human mind.

To finish on a mundane and practical note, it has been my experience that many students are troubled by the way in which their teachers make free use of Greek letters. They are indeed useful, and I plead guilty to using them in this book. In mitigation, I draw attention to an appendix in which the Greek letters, with their names, are listed.

---

[1] Joseph Fourier, 1768–1830.
[2] Godfrey Harold Hardy, 1877–1947.

1.2 The Concept of Number

The concept of number is fundamental to our way of life. Small children quickly grasp the concept of the natural numbers $1, 2, 3, \ldots$, and at an early age become proud of their ability to count up to 20, or 100, or whatever. They soon also grasp the concept of fractions, and woe betide the parent who fails to divide a piece of cake into two halves with sufficient accuracy. Negative numbers come later, but one soon understands the difference between profit and loss, between early and late, and it is in concepts such as these that negative numbers find a natural place.

Unhappily the **real numbers**, central to the topic of this book, are a great deal more sophisticated. It is certainly both natural and useful to associate positive numbers with lengths, areas and volumes. This is an idea that goes back at least to the ancient Greeks, and we use it whenever we draw a graph, a histogram, or almost any kind of chart. However, as was realised a very long time ago, this association leads to a difficulty, for not every length is expressible as a fraction. To take the simplest possible case, if we have a triangle ABC,

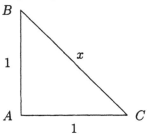

right-angled at A, and with AB and AC of length 1, then by the celebrated Theorem of Pythagoras the length of BC is a number x such that

$$x^2 = 1^2 + 1^2 = 2.$$

Let us suppose that x is expressible as a fraction m/n, where m and n are natural numbers. We may certainly suppose that the fraction is in its "lowest terms", by which we mean that m and n have no common factor, for we can always divide both m and n by any common factor to obtain a fraction with the same value as before. From the assumption that $(m/n)^2 = 2$ we easily deduce that

$$m^2 = 2n^2. \tag{1.1}$$

It follows that m^2 is divisible by 2, and this can happen only if m is divisible by 2. Accordingly, we can write $m = 2l$, where l is again a natural number. We can thus rewrite (1.1) as $4l^2 = 2n^2$, and deduce that

$$n^2 = 2l^2. \tag{1.2}$$

From (1.2) it follows by the same argument as before that n is divisible by 2. Thus both m and n are divisible by 2, and we have a contradiction, since we began with a fraction m/n in lowest terms. We are forced to conclude that x is not expressible as a fraction, that it is an **irrational** number. We denote it by $\sqrt{2}$.

It would be wrong to conclude that this is an isolated phenomenon. In fact we can assert that $\sqrt{n}$ is irrational for every natural number n that is not a perfect square. Let n be such a number. Then there exists a prime number p which divides n an odd number of times (for otherwise n is the square of some natural number m). Suppose that $\sqrt{n}$ is a rational number r/s, expressed in lowest terms. Then $(r/s)^2 = n$ and so

$$r^2 = ns^2. \tag{1.3}$$

We can write n as $p^{2k+1}l$, where $k \geq 0$, l is a natural number and p does not divide l. It follows that p^{2k+1} divides r^2. Now, the factorisation of r^2 into prime numbers involves only even powers of primes, and so we may deduce that p^{2k+2} divides r^2, from which it follows that p^{k+1} divides r. Thus we may write $r = p^{k+1}j$, where j is a natural number, and then (1.3) becomes

$$ls^2 = pj^2.$$

Since p divides ls^2 but does not divide l, we are forced to conclude that p divides s^2, and from this it follows that p divides s. Thus r and s have a common factor p, a contradiction.

We have established that there are infinitely many irrational numbers, but our result above is not by any means the end of the story. For example, it is not hard to show that numbers such as $\sqrt{2} + \sqrt{3}$ are irrational. (See Exercise 1.1.) Suffice it to say that a number system that is going to relate to the geometric notion of length must be rich enough to contain all the irrational numbers we have identified, and many more besides.

EXERCISES

1.1 Let x be a real number. Show that if x^2 is irrational, then so is x. Deduce that $\sqrt{2} + \sqrt{3}$ is irrational.

1.3 The Language of Set Theory

Before proceeding with a further examination of our concept of number, we pause to record some of the language and notations we shall be using. In analysis, as in other areas of mathematics, the language of set theory is useful in providing compact and precise statements of complex assertions. We take the intuitive notion of a set as a collection of objects (usually numbers) called the **elements** of the set. If A denotes a set, and if a is an element of A, we shall write $a \in A$, and read "a belongs to A". If a is *not* an element of A we write $a \notin A$ and read "a does not belong to A". We can specify a set simply by listing the elements inside a pair of curly brackets: thus

$$A = \{2, 4, 6, 8\}$$

means that A is the set whose elements are 2, 4, 6 and 8. This is obviously most useful for **finite** sets (where the number of elements is finite), but can sometimes be useful for infinite sets also: for example,

$$\mathbb{N} = \{1, 2, 3, \ldots\} \tag{1.4}$$

is the set of all **natural numbers**, and

$$E = \{2, 4, 6, \ldots\} \tag{1.5}$$

is the set of even natural numbers. Observe that we are making the convention that $\mathbb{N}$ does not include 0. Our notation for the set $\{0, 1, 2, \ldots\}$ will be $\mathbb{N}^0$.

This may be a good moment to list the standard notations for the most important sets of numbers:

$\mathbb{N}$	the set of natural numbers;
$\mathbb{Z}$	the set $\{\ldots, -2, -1, 0, 1, 2, \ldots\}$ of all integers;
$\mathbb{Q}$	the set of all rational numbers;
$\mathbb{R}$	the set of all real numbers;
$\mathbb{C}$	the set of all complex numbers.

The curly brackets notation is used also for **singleton** sets, consisting of a single element. There is a logical distinction between the number 2 and the set $\{2\}$ whose sole element is 2 – though it must be said that in most circumstances the distinction is not a very important one.

The other way of specifying a set is by means of a **defining property**. We can specify the set of even natural numbers in several different ways as

$$\{n : n \in \mathbb{N} \text{ and } n \text{ is divisible by } 2\}, \quad \{n \in \mathbb{N} : n \text{ is divisible by } 2\},$$

$$\{2n : n \in \mathbb{N}\}.$$

(Read the colon (:) as 'such that'.)

The idea of a **subset** of a set is an obvious one; for example, the set

$$B = \{n \in \mathbb{N} : n \text{ is divisible by } 4\} \tag{1.6}$$

of natural numbers divisible by 4 is a subset of the set E of even numbers defined above. So is the set

$$C = \{4n + 2 : n = 0, 1, 2, \ldots\} = \{2, 6, 10, \ldots\}. \tag{1.7}$$

We write $B \subseteq E$, $C \subseteq E$, or equivalently $E \supseteq B$, $E \supseteq C$. In general, for sets X and Y we say that Y is a **subset** of X, and write $Y \subseteq X$, if we have the implication

$$x \in Y \implies x \in X.$$

Notice that the definition means that the set X itself is a subset of X. If we wish to specify that $Y \subseteq X$ but that $Y \neq X$, then we write $Y \subset X$, and say that Y is a **proper** subset of X. Thus, with the sets defined in (1.5), (1.6) and (1.7) we can say that $B \subset E$ and $C \subset E$, since both are proper subsets.

Among the subsets of any set A is the **empty** (or **null**) set, always denoted by $\emptyset$, having no elements at all. We talk of "the" empty set, for there is only one: after all, the set consisting of no apples is indistinguishable from the set consisting of no oranges.

Given two subsets X and Y of a set A, we define $X \cup Y$, the **union** of X and Y, by

$$X \cup Y = \{a \in A : a \in X \text{ or } a \in Y\}. \tag{1.8}$$

Here "or" is to be interpreted *inclusively*. That is, we mean that $a \in X$ or $b \in X$ *or both*. Thus

$$\{1, 2, 3, 4\} \cup \{4, 5, 6, 7\} = \{1, 2, 3, 4, 5, 6, 7\}.$$

The other important operation on subsets is that of intersection: given X and Y as before, we define $X \cap Y$, the **intersection** of X and Y, by

$$X \cap Y = \{a \in A : a \in X \text{ and } a \in Y\}. \tag{1.9}$$

Thus

$$\{1, 2, 3, 4\} \cap \{4, 5, 6, 7\} = \{4\}.$$

Two subsets are said to be **disjoint** if their intersection is the empty set. Thus, for the sets E, B and C defined in (1.5), (1.6) and (1.7) above, we see that $B \cap C = \emptyset$, and so we may say that B and C are disjoint.

The next operation on subsets is that of complementation. Given a subset X of a set A, we define $A \setminus X$, the **complement** of X in A, by

$$A \setminus X = \{a \in A : a \notin X\}. \tag{1.10}$$

Notice that

$$X \cup (A \setminus X) = A, \quad X \cap (A \setminus X) = \emptyset.$$

We shall sometimes want to use the notation $A \setminus X$, meaning the set of elements in A but not in X, in cases where X is not a subset of A. Thus $\{1,2,3\} \setminus \{2,3,4\} = \{1\}$.

The three operations give rise to an algebra of subsets, usually called Boolean[3] algebra. This will not greatly concern us here, but more information can be found in [2].

The final set-theoretic construction we shall need is the Cartesian[4] product. If A and B are sets, then we define $A \times B$, the **Cartesian product** of A and B, by

$$A \times B = \{(a,b) : a \in A, \, b \in B\}. \tag{1.11}$$

This is a generalisation of Descartes' representation of points in the plane by pairs (x, y), with x, y in $\mathbb{R}$, which is so much part of our thinking that we often refer to the plane as $\mathbb{R} \times \mathbb{R}$, or just as $\mathbb{R}^2$.

EXERCISES

1.2 Let $X = \{1,2,3,4,5,6,7,8\}$, $A = \{1,3,5,7\}$ and $B = \{1,2,3,4\}$. Write down $A \cap B$, $A \cup B$, $X \setminus A$ and $X \setminus B$. Verify that

$$X \setminus (A \cap B) = (X \setminus A) \cup (X \setminus B)$$
$$X \setminus (A \cup B) = (X \setminus A) \cap (X \setminus B).$$

1.3 Let $\mathbb{P}$ denote the set of all prime numbers. Show that the sets

$$\{p \in \mathbb{P} : p \text{ divides } 437\} \text{ and } \{p \in \mathbb{P} : p \text{ divides } 493\}$$

are disjoint.

1.4 Real Numbers

The properties of the set $\mathbb{N}$ of natural numbers and the set

$$\mathbb{Z} = \{\ldots, -2, -1, 0, 1, 2, \ldots\}$$

of integers are familiar. So too are the properties of the set $\mathbb{Q}$ of rational numbers. Every rational number is expressible as a ratio m/n, where m and n are

---

[3] George Boole, 1815–1864.
[4] René Descartes, 1596–1650

integers and $n \neq 0$. The expression is not unique, since $km/kn = m/n$ for every k in $\mathbb{Z} \setminus \{0\}$. We add and multiply rational numbers by the following rules, which derive in their entirety from the arithmetic of integers:

$$\frac{m}{n} + \frac{p}{q} = \frac{mq + np}{nq}, \qquad \frac{m}{n}\frac{p}{q} = \frac{mp}{nq}.$$

We can regard the set $\mathbb{Q}$ as containing the set $\mathbb{Z}$ by identifying, for each m in $\mathbb{Z}$, the rational number $m/1$ with the integer m. Two important rational numbers are $0 \ (= 0/1 = 0/n$ for every $n \neq 0)$, and $1 \ (= 1/1 = n/n$ for every $n \neq 0)$. The set $\mathbb{Q}$ forms what is known as a **field** with respect to the operations of addition and multiplication. Informally this is a system in which one can add, subtract, multiply and (except by 0) divide. Formally, we have axioms as follows:

(F1) $(a+b)+c = a+(b+c)$ $(a, b, c \in \mathbb{Q})$ (the **associative law for addition**);

(F2) $a + b = b + a$ $(a, b \in \mathbb{N})$ (the **commutative law for addition**);

(F3) there exists 0 in $\mathbb{Q}$ such that, for every a in $\mathbb{Q}$, $a + 0 = a$;

(F4) for each a in $\mathbb{Q}$ there exists an element $-a$ in $\mathbb{Q}$ such that $a + (-a) = 0$;

(F5) $(ab)c = a(bc)$ $(a, b, c \in \mathbb{Q})$ (the **associative law for multiplication**);

(F6) $ab = ba$ $(a, b \in \mathbb{Q})$ (the **commutative law for multiplication**);

(F7) there exists 1 in $\mathbb{Q}$ such that, for every a in $\mathbb{Q}$, $a1 = a$;

(F8) for all $a \neq 0$ in $\mathbb{Q}$, there exists $1/a$ in $\mathbb{Q}$ such that $a(1/a) = 1$ (the **law of the reciprocal**);

(F9) $a(b + c) = ab + ac$ $(a, b, c \in \mathbb{Q})$ (the **distributive law**).

The negative of m/n is $(-m)/n$, and the reciprocal of m/n is n/m.

In addition to the algebraic operations of addition and multiplication, $\mathbb{Q}$ has a natural order relation $<$, inherited from the obvious order relation

$$\cdots < -2 < -1 < 0 < 1 < 2 < \cdots$$

on $\mathbb{Z}$. Given m/n and p/q we can certainly make sure that the fractions are expressed in such a way that n and q are positive, and then define

$$\frac{m}{n} < \frac{p}{q} \quad \text{if} \quad mq < np.$$

In the standard way we can choose to write $a > b$ (where $a, b \in \mathbb{Q}$) as an alternative to $b < a$. Also we can use the notations $a \leq b$ to mean "$a < b$ or $a = b$", and $a \geq b$ to mean "$a > b$ or $a = b$".

The axioms relating to the order relation are:

(O1) $a < b$ and $b < c$ implies $a < c$ $(a, b, c \in \mathbb{Q})$ (the **transitive law**);

(O2) for all a, b in $\mathbb{Q}$, exactly one of the following holds: $a < b$, or $a > b$, or $a = b$ (the **trichotomy law**);

(O3) for all a, b, c in $\mathbb{Q}$, $a < b$ implies $a + c < b + c$ (the law of **compatibility with addition**);

(O4) for all a, b, c in $\mathbb{Q}$, if $a < b$ and $c > 0$, then $ac < bc$ (the law of **compatibility with multiplication**).

The axioms (F1)–(F9) and (O1)–(O4) make $\mathbb{Q}$ into what is called an **ordered field**, and are the basis of the algebra one learns in secondary school.

It is easy to see that $\mathbb{Q}$ has the property of **density**: between any two rational numbers q and r (with $q < r$) there is a rational number s such that $q < s < r$: simply take $s = (q + r)/2$. The density property implies in fact that there are *infinitely many* rational numbers between q and r; for we have

$$q < (q + s)/2 < s < (s + r)/2 < r,$$

and we can go on subdividing indefinitely.

It is clear also that $\mathbb{Q}$ has the so-called **archimedean property** [5], that

$$\text{for all } q > 0 \text{ in } \mathbb{Q} \text{ there exists } n \text{ in } \mathbb{N} \text{ such that } n > q: \qquad (1.12)$$

for $q = r/s$ with r, s in $\mathbb{N}$, simply take $n = 2r$. The property implies that for any two positive rational numbers q and r there exists a natural number n such that $nq > r$. Simply choose n such that $n > r/q$.

The set of real numbers, which we will always denote by $\mathbb{R}$, and which contains $\mathbb{Q}$, is also an ordered field. The density property of $\mathbb{Q}$ might appear to leave no gaps for irrational numbers, but if it is to be of any use our number system must squeeze them in somehow. We now describe the "completeness" property that distinguishes $\mathbb{R}$ from $\mathbb{Q}$.

Let us look first at $\mathbb{Q}$. A subset B of $\mathbb{Q}$ is said to be **bounded above** if there exists K in $\mathbb{Q}$ such that $K \geq b$ for all b in B; such a K is called an **upper bound** of B. Thus, for example, the set $\{a \in \mathbb{Q} : a < 4\}$ is bounded above by 4. (It is also bounded above by 5, or indeed by any number greater than 4.) By contrast, the subset $\mathbb{N}$ of $\mathbb{Q}$ is not bounded above. Among the upper bounds of the set $\{a \in \mathbb{Q} : a < 4\}$ there is a *least* upper bound, namely 4, for no number strictly less than 4 will be an upper bound of the set. Formally, if B is a non-empty subset of $\mathbb{Q}$ which is bounded above, then a **least upper bound** of B is a number k in $\mathbb{Q}$ such that (i) k is an upper bound of B; (ii) if c is an upper bound for B, then $c \geq k$.

It is easy to see that if such a number exists then it is unique; for if we suppose that k_1, k_2 are both least upper bounds of B, then $k_1 \geq k_2$, since k_1

---

[5] Archimedes of Syracuse, 287–212 BC

is an upper bound and k_2 is least, and equally $k_2 \geq k_1$, since k_2 is an upper bound and k_1 is least. So if the number exists we may reasonably call it **the** least upper bound.

However, within the set $\mathbb{Q}$, it may fail to exist. Consider the subset

$$C = \{a \in \mathbb{Q} : a \geq 0 \text{ and } a^2 < 2\}.$$

It is certainly bounded above: if $a \in C$ then certainly $a^2 < 4$ and so $a < 2$. The upper bound 2 is certainly not the least upper bound, for we can see that the smaller number 3/2 is also an upper bound: if $a \in C$ then $a^2 < 9/4$ and so $a < 3/2$. But that is not least either. In fact, if we take the approximation 1.414213562 to $\sqrt{2}$ given by a calculator, we can see that 1.5 (already considered), 1.42, 1.415, 1.4143, 1.41422, 1.41214, ... give a sequence of ever smaller rational numbers all of which are upper bounds of C. It is precisely because $\sqrt{2}$ is irrational that we can never find a least upper bound within the field of rational numbers. We can find an *irrational* number that is a least upper bound of C, namely $\sqrt{2}$, and it is this observation that prompts the following definition.

The set $\mathbb{R}$ of **real numbers** is an ordered field satisfying

Property 1.1 (The Axiom of Archimedes)

For all $x > 0$ in $\mathbb{R}$ there exists n in $\mathbb{N}$ such that $n > x$,

and

Property 1.2 (The Axiom of Completeness)

Every non-empty subset of $\mathbb{R}$ that is bounded above has a least upper bound in $\mathbb{R}$.

Of course it is legitimate to ask whether $\mathbb{R}$ can be shown to exist. It can, provided we make the very reasonable assumption that the set $\mathbb{N}$ of natural numbers exists – and it was this realisation that prompted the famous remark by Kronecker[6] that "God created the integers, all else is the work of man." [7] The construction process is, however, lengthy and tedious, and in a first serious course on real analysis it is sufficient to "believe" that the numbers on the real line have the properties mentioned, including the crucial property (1.2).

What about sets that are bounded below? The definitions of "bounded below", "lower bound" and "greatest lower bound" are clear, and the one-

---

[6] Leopold Kronecker, 1823–1891
[7] Die ganzen Zahlen hat der liebe Gott gemacht, alles andere ist Menschenwerk.

sided appearance of the axiom (1.2) is only apparent. Indeed it *follows* from the Axiom of Completeness that we also have the property

Theorem 1.3

Every non-empty subset of $\mathbb{R}$ that is bounded below has a greatest lower bound in $\mathbb{R}$.

Proof

Let A be a non-empty subset of $\mathbb{R}$ that is bounded below, in the obvious sense that there exists k in $\mathbb{R}$ such that $k \leq a$ for every a in A. If we now consider the set $-A = \{-a : a \in A\}$, we see that $-k$ is an upper bound for $-A$. By the axiom (1.2), $-A$ has a least upper bound c, and from this it follows that $-c$ is a greatest lower bound of A. $\qquad\square$

It is useful to have compact notation for some particularly important subsets of the field $\mathbb{R}$ of real numbers, called **intervals**. Let a, b be real numbers, with $a < b$. Then:

$$\{x \in \mathbb{R} : a \leq x \leq b\}, \text{ denoted by } [a, b];$$
$$\{x \in \mathbb{R} : a \leq x < b\}, \text{ denoted by } [a, b);$$
$$\{x \in \mathbb{R} : a < x \leq b\}, \text{ denoted by } (a, b];$$
$$\{x \in \mathbb{R} : a < x < b\}, \text{ denoted by } (a, b);$$
$$\{x \in \mathbb{R} : x \geq a\}, \text{ denoted by } [a, \infty);$$
$$\{x \in \mathbb{R} : x > a\}, \text{ denoted by } (a, \infty);$$
$$\{x \in \mathbb{R} : x \leq b\}, \text{ denoted by } (-\infty, b];$$
$$\{x \in \mathbb{R} : x < b\}, \text{ denoted by } (-\infty, b).$$

The interval $[a, b]$ is called a **closed** interval, and the interval (a, b) is called an **open** interval. It is important to realise that in using a notation such as $[a, \infty)$ we are not proclaiming a belief that there is any such real number as ∞. There is not. The meaning of the notation is precisely as we have defined it, no more, no less.

Consider a non-empty subset A of $\mathbb{R}$, and suppose that A is bounded above. From the axiom of completeness we know that A has a least upper bound. This is always referred to as the **supremum** of A, and is written $\sup A$. Similarly, if A is bounded below, the greatest lower bound of A is called the **infimum** of A and is written $\inf A$. For intervals the situation is clear:

$$\sup(a, b) = \sup[a, b] = b, \quad \inf(a, b) = \inf[a, b] = a.$$

For other sets, as we shall see in due course, it may be harder to find the supremum, but it can certainly be a useful starting point to establish that it exists.

EXERCISES

1.4 Which of the following statements are true? If the statement is true, prove it; if not, give a counterexample. Let x and y be real numbers.

a) If x is rational and y is irrational, then $x + y$ is irrational.

b) If x is rational and y is irrational, then xy is irrational.

c) If x and y are irrational, then so is $x + y$.

d) If x and y are irrational, then so is xy.

e) If x and y are irrational, then $x + y$ is rational.

f) If x and y are irrational, then xy is rational.

1.5 Let x, y be real numbers, with $x < y$. Show that, if x and y are rational, then there exists an irrational number u such that $x < u < y$.

1.6 Let $x, y \in \mathbb{R}$, with $x < y$. Show that there exists a rational number q such that $x < q < y$.

1.5 Induction

At various stages in this book we shall need to prove results by induction, and it may be necessary to remind some readers of what is involved in this process.

Let $\mathbb{P}(n)$ be a proposition concerning a natural number n.

The Principle of Induction. If $\mathbb{P}(1)$ is true and if, for all $n \geq 1$, $\mathbb{P}(n)$ implies $\mathbb{P}(n + 1)$, then $\mathbb{P}(n)$ is true for all $n \geq 1$.

Example 1.4

Show that, for all $n \geq 1$,

$$\sum_{r=1}^{n} r(r + 1) = \frac{1}{3}n(n + 1)(n + 2). \tag{1.13}$$

Solution

Denote the proposition (1.13) by $\mathbb{P}(n)$. If $n = 1$ then both the left hand side and the right hand side of the formula (1.13) take the value 2, and so $\mathbb{P}(1)$ is true. Let $n \geq 1$, and suppose that $\mathbb{P}(n)$ is true. Then

$$\sum_{r=1}^{n+1} r(r+1) = \sum_{r=1}^{n} r(r+1) + (n+1)(n+2)$$
$$= \frac{1}{3}n(n+1)(n+2) + (n+1)(n+2)$$
$$\text{(by the hypothesis that } \mathbb{P}(n) \text{ is true)}$$
$$= \frac{1}{3}(n+1)(n+2)(n+3),$$

which is $\mathbb{P}(n+1)$. We have shown that, for all $n \geq 1$, $\mathbb{P}(n)$ implies $\mathbb{P}(n+1)$, and so, by the principle of induction, $\mathbb{P}(n)$ is true for all $n \geq 1$. $\qquad\square$

Example 1.5

Into how many regions do n straight lines "in general position" divide the plane?

Solution

By "in general position" we mean that no two lines are parallel and no three lines are concurrent.

With a bit of experimentation we can discover that the number we want is $\Delta(n)$, where the first few values are given by the table

n	1	2	3	4	5
$\Delta(n)$	2	4	7	11	16

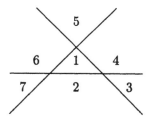

This example illustrates the disadvantage of the inductive method, for it is now necessary to make an inspired guess as to the correct formula for $\Delta(n)$. Certainly the formula

$$\Delta(n) = \frac{1}{2}(n^2 + n + 2)$$

is correct for $n = 1, 2, 3, 4, 5$. Suppose inductively that it is true for n, so that n lines divide the plane into $\frac{1}{2}(n^2 + n + 2)$ regions. The $(n+1)$th line intersects with each of the existing n lines in n points, and is divided into $n+1$ segments by these points. Each of these segments divides an existing region into two, and so the number of new regions created is $n+1$. Thus

$$\Delta(n+1) = \frac{1}{2}(n^2 + n + 2) + (n+1)$$
$$= \frac{1}{2}(n^2 + n + 2 + 2n + 2) = \frac{1}{2}[(n^2 + 2n + 1) + (n+1) + 2]$$
$$= \frac{1}{2}[(n+1)^2 + (n+1) + 2],$$

and so the result is proved by induction.   □

Remark 1.6

If inspiration fails in this example, there are ways to aid it. If one suspects that the answer is a quadratic function of n, of the form $an^2 + bn + c$, then one can put $n = 1, 2, 3$ in succession and solve the equations

$$a + b + c = 2$$
$$4a + 2b + c = 4$$
$$9a + 3b + c = 7$$

to find $a = b = 1/2$, $c = 1$.

The most substantial use we shall make of the inductive principle is in establishing the **Binomial Theorem**. We define, for integers $n \geq r \geq 0$, the **binomial coefficient** $\binom{n}{r}$ by

$$\binom{n}{r} = \frac{n(n-1)\ldots(n-r+1)}{r!} \quad \left(= \frac{n!}{r!(n-r)!}\right),$$

and observe that

$$\binom{n}{0} = \binom{n}{n} = 1, \quad \binom{n}{1} = \binom{n}{n-1} = n, \quad \binom{n}{r} = \binom{n}{n-r}.$$

The binomial theorem is as follows:

Theorem 1.7 (The Binomial Theorem)

For all a, b in $\mathbb{R}$ and all natural numbers n,

$$(a+b)^n = \sum_{r=0}^{n} \binom{n}{r} a^{n-r} b^r$$

$$= a^n + na^{n-1}b + \cdots + \binom{n}{r} a^{n-r} b^r + \cdots + nab^{n-1} + b^n.$$

Proof

This is certainly true for $n = 1$, since

$$(a+b)^1 = a + b = \binom{1}{0} a + \binom{1}{1} b.$$

Let $n \geq 1$, and suppose inductively that the theorem is true for n. To show that it is true for $n+1$ we require the identity

$$\binom{n}{r-1} + \binom{n}{r} = \binom{n+1}{r} \quad (n \geq r \geq 1). \tag{1.14}$$

This is usually called the **Pascal[8] Triangle Identity**, since in the triangle of numbers

$$\begin{array}{ccccccccc}
& & & & 1 & & 1 & & \\
& & & 1 & & 2 & & 1 & \\
& & 1 & & 3 & & 3 & & 1 \\
& 1 & & \textcircled{4} & & \textcircled{6} & & 4 & & 1 \\
1 & & 5 & & \textcircled{10} & & 10 & & 5 & & 1
\end{array}$$

each entry is the sum of the two entries above it, and since the nth row turns out to be $\binom{n}{0} \binom{n}{1} \cdots \binom{n}{n}$. Thus, if $n = 4$ and $r = 2$,

$$\binom{4}{1} + \binom{4}{2} = 4 + 6 = 10 = \binom{5}{2}.$$

The identity is easily proved:

$$\binom{n}{r-1} + \binom{n}{r} = \frac{n(n-1)\ldots(n-r+2)}{(r-1)!} + \frac{n(n-1)\ldots(n-r+1)}{r!}$$

$$= \frac{n(n-1)\ldots(n-r+2)}{r!} [r + (n-r+1)]$$

---

[8] Blaise Pascal, 1623–1662

$$= \frac{(n+1)[(n+1)-1]\dots[(n+1)-r+1]}{r!}$$

$$= \binom{n+1}{r}.$$

For the induction step,

$$(a+b)^{n+1} = (a+b)(a+b)^n = (a+b)\left(\sum_{r=0}^{n}\binom{n}{r}a^{n-r}b^r\right)$$

$$= (a+b)\left(a^n + \dots + \binom{n}{r-1}a^{n-r+1}b^{r-1} + \binom{n}{r}a^{n-r}b^r + \dots + b^n\right).$$

For $r = 1, 2, \dots, n$, the term in $a^{(n+1)-r}b^r$ is

$$b\cdot\binom{n}{r-1}a^{n-r+1}b^{r-1} + a\cdot\binom{n}{r}a^{n-r}b^r,$$

and so the coefficient is

$$\binom{n}{r-1} + \binom{n}{r} = \binom{n+1}{r}.$$

The coefficients of a^{n+1} and b^{n+1} are both 1, and so we conclude that

$$(a+b)^{n+1} = \sum_{r=0}^{n+1}\binom{n+1}{r}a^{(n+1)-r}b^r.$$

Hence, by induction, the result is true for all $n \geq 1$. □

There is a second version of the induction principle as follows:

The Second Principle of Induction. If $\mathbb{P}(1)$ is true and if, for all $n \geq 1$, the truth of $\mathbb{P}(k)$ for all $k < n$ implies that $\mathbb{P}(n)$ is true, then $\mathbb{P}(n)$ is true for all $n \geq 1$.

Example 1.8

Suppose that the sequence of numbers $a_1, a_2, a_3, \dots$ is defined by

$$a_1 = 1, a_2 = 3, \quad a_n = 2a_{n-1} + a_{n-2} \ (n \geq 3).$$

Prove that, for all $n \geq 1$

$$a_n = \frac{1}{2}[(1+\sqrt{2})^n + (1-\sqrt{2})^n]. \tag{1.15}$$

Solution

In this example logic requires that we "anchor" the induction by verifying that the result is true for $n = 1$ and $n = 2$. This is easily done:

$$\frac{1}{2}[(1 + \sqrt{2}) + (1 - \sqrt{2})] = 1 = a_1 ,$$

$$\frac{1}{2}[(1 + \sqrt{2})^2 + (1 - \sqrt{2})^2] = \frac{1}{2}[(3 + 2\sqrt{2}) + (3 - 2\sqrt{2})] = 3 = a_2 .$$

If we now assume that $n \geq 3$ and that $\mathbb{P}(k)$, given by

$$a_k = \frac{1}{2}[(1 + \sqrt{2})^k + (1 - \sqrt{2})^k] ,$$

holds for all $k < n$, we see that

$$
\begin{aligned}
a_n &= 2a_{n-1} + a_{n-2} \\
&= [(1 + \sqrt{2})^{n-1} + (1 - \sqrt{2})^{n-1}] + \frac{1}{2}[(1 + \sqrt{2})^{n-2} + (1 - \sqrt{2})^{n-2}] \\
&= \frac{1}{2}[(1 + \sqrt{2})^{n-2}(2 + 2\sqrt{2} + 1) + (1 - \sqrt{2})^{n-2}(2 - 2\sqrt{2} + 1)] \\
&= \frac{1}{2}[(1 + \sqrt{2})^{n-2}(1 + \sqrt{2})^2 + (1 - \sqrt{2})^{n-2}(1 - \sqrt{2})^2] \\
&= \frac{1}{2}[(1 + \sqrt{2})^n + (1 - \sqrt{2})^n] .
\end{aligned}
$$

Hence the result holds for all $n \geq 1$. $\qquad\square$

EXERCISES

1.7 Show that

$$1^2 + 2^2 + \cdots + n^2 = \frac{1}{6}n(n + 1)(2n + 1) .$$

1.8 Show that

$$1^3 + 2^3 + \cdots + n^3 = \frac{1}{4}n^2(n + 1)^2 .$$

1.9 Prove by induction the formula for the sum of an **arithmetic series**:

$$a + (a + d) + (a + 2d) + \cdots + (a + (n - 1)d) = na + \frac{1}{2}n(n - 1)d .$$

1.10 Prove by induction the formula for the sum of a **geometric series**:

$$a + ar + ar^2 + \cdots + ar^{n-1} = \frac{a(1 - r^n)}{1 - r} \quad (r \neq 1) .$$

1.11 Prove by induction that, for all $x \neq 1$,

$$1 + 2x + 3x^2 + \cdots + nx^{n-1} = \frac{nx^{n+1} - (n+1)x^n + 1}{(x-1)^2}.$$

1.12 Show that $(n+1)^4 < 4n^4$ whenever $n \geq 3$. Hence show by induction that $4^n > n^4$ for all $n \geq 5$.

1.13 Let the numbers $q_1, q_2, q_3, \ldots$ be defined by

$$q_1 = 2, \quad q_n = 3q_{n-1} - 1 \; (n \geq 2).$$

Show by induction that, for all $n \geq 1$,

$$q_n = \frac{1}{2}(3^n + 1).$$

1.14 Let the numbers $a_0, a_1, a_2, \ldots$ be defined by

$$a_0 = 1, \, a_1 = 3 \quad a_n = 4(a_{n-1} - a_{n-2}) \; (n \geq 2).$$

Show by induction that $a_n = 2^{n-1}(n+2)$ for all $n \geq 0$.

1.6 Inequalities

The manipulation of inequalities (statements involving the symbols $<$, $>$, $\leq$, $\geq$) is at the heart of real analysis, and in this short section we examine some of the crucial notations and techniques involved. If x is a real number then $|x|$, the **absolute magnitude**, or **modulus**, of x, is defined as follows:

$$|x| = \left\{ \begin{array}{ll} x & \text{if } x \geq 0 \\ -x & \text{if } x < 0. \end{array} \right. \tag{1.16}$$

Thus, for example, $|3| = 3$, $|-4| = -(-4) = 4$. Notice that $|x| \geq 0$ for every x in $\mathbb{R}$, and that $|x| = 0$ if and only if $x = 0$. It is sometimes useful to note that, for every real number x,

$$\sqrt{x^2} = |x|. \tag{1.17}$$

Note here that we always mean the *positive* square root when we write $\sqrt{}$; thus, for example

$$\sqrt{(-3)^2} = \sqrt{9} = 3 = |-3|.$$

We collect together some of the most important properties in a theorem as follows:

Theorem 1.9

Let x, y be real numbers.

(i) $|-x| = |x|$, $x \le |x|$, $|x| = \max\{x, -x\}$;

(ii) $|xy| = |x||y|$;

(iii) $|x + y| \le |x| + |y|$;

(iv) $\big||x| - |y|\big| \le |x - y|$.

Proof

(i) These follow immediately from the definition of $|x|$.

(ii) The result is immediate if $x = 0$ or $y = 0$. If x and y are both positive, then so is xy, and so
$$|xy| = xy = |x||y|.$$
If x and y are both negative, then xy is positive, and
$$|xy| = xy = (-x)(-y) = |x||y|.$$
If x is positive and y is negative, then xy is negative, and
$$|xy| = -xy = x(-y) = |x||y|.$$
The final case, where x is negative and y positive, is similar.

(iii) This can also be proved by considering cases. Alternatively, we can observe that, for all x, y in $\mathbb{R}$,
$$\begin{aligned}
(x + y)^2 = x^2 + 2xy + y^2 &= |x|^2 + 2xy + |y|^2 \\
&\le |x|^2 + 2|x||y| + |y|^2 \quad \text{(by parts (i) and (ii))} \\
&= (|x| + |y|)^2.
\end{aligned}$$
Then, taking (positive) square roots, and using the observation (1.17), we have
$$|x + y| \le |x| + |y|,$$
as required.

(iv) From the obvious equality $x = (x - y) + y$ we deduce from part (iii) that
$$|x| = |(x - y) + y| \le |x - y| + |y|;$$
hence
$$|x| - |y| \le |x - y|. \tag{1.18}$$

Similarly, from the observation that $y = (y - x) + x$ we deduce that

$$|y| \leq |y - x| + |x| = |x - y| + |x|;$$

hence

$$|y| - |x| \leq |x - y|. \tag{1.19}$$

From (1.18) and (1.19) we deduce that

$$\max\left\{|x| - |y|, \ |y| - |x|\right\} \leq |x - y|$$

and hence, using part (i), that

$$\Big||x| - |y|\Big| \leq |x - y|,$$

as required. $\square$

Remark 1.10

The inequalities in parts (iii) and (iv) may well be proper: for example,

$$|3 + (-2)| < |3| + |-2|, \quad \Big||3| - |-2|\Big| < |3 - (-2)|.$$

Remark 1.11

From part (ii) we can immediately deduce that

$$|x - y| \leq |x| + |y|, \tag{1.20}$$

for

$$|x - y| = |x + (-y)| \leq |x| + |-y| = |x| + |y|.$$

Remark 1.12

From (ii) it also follows that

$$|x + y + z| = |(x + y) + z| \leq |x + y| + |z| \leq (|x| + |y|) + |z|.$$

More generally, we have that for all real numbers $x_1, x_2, \ldots, x_n$,

$$|x_1 + x_2 + \cdots + x_n| \leq |x_1| + |x_2| + \cdots + |x_n|. \tag{1.21}$$

The manipulation of inequalities is a crucial technique in analysis. In addition to the axioms (O1) to (O4) on page 8 certain key pieces of information, all of which follow from the axioms, are worth listing here (x and y are real numbers):

$$\text{if } x < y \text{ then } -x > -y; \tag{1.22}$$

$$\text{if } 0 < x < y \text{ then } \tfrac{1}{x} > \tfrac{1}{y}; \tag{1.23}$$

$$x^2 \geq 0 \text{ for all } x \text{ in } \mathbb{R}, \text{ and } x^2 = 0 \text{ if and only if } x = 0. \tag{1.24}$$

The property (1.24) certainly has far-reaching consequences. First, we record some of the crucial properties of the quadratic expression $ax^2 + bx + c$, where $a \neq 0$.

Theorem 1.13

Let $a, b, c \in \mathbb{R}$, with $a \neq 0$.

(i) $ax^2 + bx + c \geq 0$ for all x in R if and only if $a > 0$ and $b^2 - 4ac \leq 0$;

(ii) the equation $ax^2 + bx + c = 0$ has real roots if and only if $b^2 - 4ac \geq 0$.

(iii) if $a > 0$, and if the equation $ax^2 + bx + c = 0$ has real roots α and β, where $\alpha < \beta$, then

$$\{x \in \mathbb{R} : ax^2 + bx + c \leq 0\} = [\alpha, \beta].$$

Proof

The crucial observation that gives all three assertions is as follows:

$$ax^2 + bx + c = a\left(x^2 + \frac{b}{a}x + \frac{c}{a}\right) = a\left(x^2 + \frac{b}{a}x + \frac{b^2}{4a^2} + \frac{c}{a} - \frac{b^2}{4a^2}\right)$$

$$= a\left[\left(x + \frac{b}{2a}\right)^2 - \frac{b^2 - 4ac}{4a^2}\right].$$

(This useful technique is known as **completing the square**.) If $a > 0$ and $b^2 - 4ac \leq 0$ then both terms inside the square brackets are non-negative, and so $ax^2 + bx + c \geq 0$ for all x. Similarly, if $a < 0$ and $b^2 - 4ac < 0$ then $ax^2 + bx + c < 0$ for all x. On the other hand, if $b^2 - 4ac > 0$ then the expression inside the square brackets changes sign when

$$\left(x + \frac{b}{2a}\right)^2 = \frac{b^2 - 4ac}{4a^2},$$

that is, when

$$x = \frac{-b \pm \sqrt{b^2 - 4ac}}{2a}.$$

These are the **roots** of the quadratic equation $ax^2 + bx + c = 0$. If $a > 0$ we have $\alpha < \beta$, where

$$\alpha = \frac{-b - \sqrt{b^2 - 4ac}}{2a}, \quad \beta = \frac{-b + \sqrt{b^2 - 4ac}}{2a}$$

and we can write

$$ax^2 + bx + c = a(x - \alpha)(x - \beta).$$

The sign of $a(x - \alpha)(x - \beta)$ changes in accordance with the following table:

x	$(-\infty, \alpha)$	α	(α, β)	β	(β, ∞)
$x - \alpha$	$-$	0	$+$	$+$	$+$
$x - \beta$	$-$	$-$	$-$	0	$+$
$ax^2 + bx + c$	$+$	0	$-$	0	$+$

If $b^2 - 4ac = 0$, then $ax^2 + bx + c = 0$ if and only if $x = -b/2a$. $\qquad\square$

It is helpful to remember the properties of $ax^2 + bx + c$ in graphical form. As we have seen, the nature of the graph is determined by the sign of a and the sign of the **discriminant** $\Delta = b^2 - 4ac$. Figures 1.1 and 1.2 show typical graphs in the cases (from left to right) $\Delta < 0$, $\Delta > 0$, $\Delta = 0$.

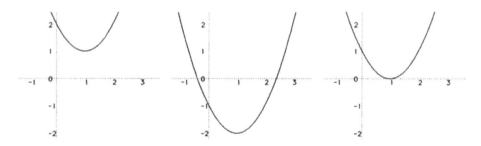

Figure 1.1. Quadratic graphs: $a > 0$

The important **Cauchy**[9]–**Schwarz**[10] inequality

$$(a_1 b_1 + a_2 b_2 + \cdots + a_n b_n)^2 \le (a_1^2 + a_2^2 + \cdots + a_n^2)(b_1^2 + b_2^2 + \cdots + b_n^2) \quad (1.25)$$

[9] Augustin-Louis Cauchy, 1789–1857
[10] Karl Hermann Amandus Schwarz, 1843–1921

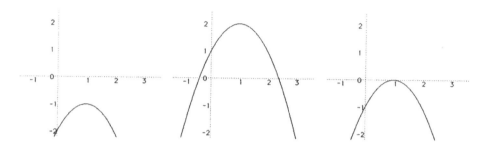

Figure 1.2. Quadratic graphs: $a < 0$

is in essence a consequence of these observations. If $n = 2$ we can easily see that

$$(a_1^2 + a_2^2)(b_1^2 + b_2^2) - (a_1b_1 + a_2b_2)^2$$
$$= a_1^2b_1^2 + a_1^2b_2^2 + a_2^2b_1^2 + a_2^2b_2^2 - a_1^2b_1^2 - a_2^2b_2^2 - 2a_1b_1a_2b_2$$
$$= a_1^2b_2^2 - 2a_1b_1a_2b_2 + a_2^2b_1^2 = (a_1b_2 - a_2b_1)^2 \geq 0.$$

For $n \geq 3$ the algebra becomes more complicated, though still manageable, but a more oblique approach is effective. For any real number λ,

$$(\lambda a_i + b_i)^2 \geq 0 \quad (i = 1, 2, \ldots, n),$$

and so

$$\sum_{i=1}^{n} (\lambda a_i + b_i)^2 \geq 0.$$

If we write each $(\lambda a_i + b_i)^2$ as $\lambda^2 a_i^2 + 2\lambda a_i b_i + b_i^2$ and regroup the terms, we see that

$$A\lambda^2 + 2B\lambda + C \geq 0$$

for all λ, where

$$A = \sum_{i=1}^{n} a_i^2, \quad C = \sum_{i=1}^{n} b_i^2, \quad B = \sum_{i=1}^{n} a_i b_i.$$

Now, we may safely assume that $A > 0$, for otherwise we have $a_1 = a_2 = \cdots = a_n = 0$, and the inequality (1.25) is trivially true. Hence, by Theorem 1.13(i), $(2B)^2 - 4AC \leq 0$, and so $B^2 \leq AC$, exactly as required.

Remark 1.14

The Cauchy–Schwarz inequality can be expressed compactly in vector notation. If $\mathbf{a} = (a_1, a_2, \ldots, a_n)$, $\mathbf{b} = (b_1, b_2, \ldots, b_n)$, if $\mathbf{a.b}$ is the **scalar product** $a_1b_1 +$

$a_2 b_2 + \cdots + a_n b_n$ of $\mathbf{a}$ and $\mathbf{b}$, and if $\|\mathbf{a}\|$ is defined as $\sqrt{a_1^2 + a_2^2 + \cdots + a_n^2}$, then the inequality states that

$$(\mathbf{a}.\mathbf{b})^2 \leq \|\mathbf{a}\|^2 \|\mathbf{b}\|^2.$$

Another fundamental inequality is also an immediate consequence of the property (1.24). If a, b are positive real numbers then the **arithmetic mean** of a and b is defined to be $(a+b)/2$, and the **geometric mean**[11] of a and b is defined as $\sqrt{ab}$. The **arithmetic-geometric inequality** states that, for all $a, b \geq 0$,

$$\sqrt{ab} \leq (a+b)/2. \qquad (1.26)$$

This is easily proved: since $(a-b)^2 \geq 0$ for all a, b, we have

$$(a+b)^2 - 4ab = a^2 - 2ab - b^2 = (a-b)^2 \geq 0.$$

Thus

$$ab \leq \frac{(a+b)^2}{4},$$

and so, taking positive square roots, we have $\sqrt{ab} \leq (a+b)/2$, as required.

We end this chapter with an inequality that does *not* depend on the positivity of squares. The factorisation

$$x^n - 1 = (x-1)(x^{n-1} + x^{n-2} + \cdots + x + 1) \qquad (1.27)$$

is valid for every integer $n \geq 2$. If $x > 1$ then

$$x^{n-1} + x^{n-2} + \cdots + x + 1 > n,$$

and so

$$x^n - 1 > n(x-1) \quad (x > 1,\ n \geq 2). \qquad (1.28)$$

Example 1.15

Describe the set

$$S = \left\{ x \in \mathbb{R} : \frac{x+1}{2x-3} \leq 1 \right\}.$$

---

[11] The terms arise because the numbers $a, (a+b)/2, b$ are in arithmetic progression (that is, $(a+b)/2 - a = b - (a+b)/2$), while the terms $a, \sqrt{ab}, b$ are in geometric progression (that is, $\sqrt{ab}/a = b/\sqrt{ab}$).

Solution

$$\frac{x+1}{2x-3} \leq 1 \text{ if and only if } 1 - \frac{x+1}{2x-3} \geq 0,$$

$$\text{that is, if and only if } \frac{2x-3-x-1}{2x-3} = \frac{x-4}{2x-3} \geq 0.$$

Here we have the table

x	$(-\infty, \frac{3}{2})$	$\frac{3}{2}$	$(\frac{3}{2}, 4)$	4	$(4, \infty)$
$2x-3$	$-$	0	$+$	$+$	$+$
$x-4$	$-$	$-$	$-$	0	$+$
$(x-4)/(2x-3)$	$+$	∞	$-$	0	$+$

(The entry ∞ in the last row of the table indicates that the expression $(x-4)/(2x-3)$ is undefined when $x = 3/2$.) So $S = (-\infty, 3/2) \cup [4, \infty)$. Notice that it is incorrect to write

$$\frac{x+1}{2x-3} \leq 1 \text{ if and only if } x + 1 \leq 2x - 3,$$

since we do not know that $2x - 3$ is positive, and so multiplying by $2x - 3$ may *reverse* the inequality. The above analysis of the factors is just the same as for the quadratic expression $(x-4)(2x-3)$, except that the quotient $(x-4)/(2x-3)$ is undefined for $x = 3/2$. $\qquad\square$

EXERCISES

1.15 Prove the following statements concerning real numbers.

a) $x^2 + 4x + 5 > 0$.

b) $x^2 + 5xy + 7y^2 \geq 0$.

c) For non-zero a, b and c,

$$a^2 + b^2 + c^2 + \frac{1}{a^2} + \frac{1}{b^2} + \frac{1}{c^2} \geq 6;$$

show also that equality holds if and only if $a, b, c \in \{-1, 1\}$.

1.16 Let $0 < a \leq b$. The **harmonic mean** of a and b is defined by

$$H = \frac{2ab}{a+b}.$$

Show that $a \leq H \leq b$. Denote the geometric mean by G. Show that $H \leq G$, with equality if and only if $a = b$.

1.17 Show that the following two sets are equal:

$$A = \{(x, y) \in \mathbb{R} \times \mathbb{R} : x^2 + y^2 + 2x - 4y + 6 = 0\},$$
$$B = \{(x, y) \in \mathbb{R} \times \mathbb{R} : x^2 + 4xy + 5y^2 + 2x + 4y + 2 = 0\}.$$

1.18 Let x, a and δ be real numbers. Show that the statements $|x - a| < \delta$ and $a - \delta < x < a + \delta$ are equivalent.

1.19 Show that, if $a > 0$ and $0 < r < 1$, then, for all $n \geq 1$

$$a + ar + ar^2 + \cdots + ar^n < \frac{a}{1 - r}.$$

1.20 Show that, if $0 < |x_2| < |x_1|$ then, for all $n \geq 1$,

$$n \left| \frac{x_2}{x_1} \right|^{n-1} < \frac{|x_1|}{|x_1| - |x_2|}.$$

1.21 Express the set

$$\left\{ x \in \mathbb{R} : \frac{3x + 2}{x - 1} < 1 \right\}$$

as an interval.

1.22 Show that, for all x, y in $\mathbb{R}$,

$$\max\{x, y\} = \frac{1}{2}(x + y + |x - y|),$$
$$\min\{x, y\} = \frac{1}{2}(x + y - |x - y|).$$

1.23 Let a, b c and d be real numbers. Show that $ab - cd = b(a-c) + c(b-d)$, and deduce that

$$|ab - cd| \leq |b||a - c| + |c||b - d|.$$

Let ϵ, K and L be positive real numbers, and suppose that $|b| < K$, $|c| < L$, $|a - c| < \epsilon/(2K)$, $|b - d| < \epsilon/(2L)$. Deduce that $|ab - cd| < \epsilon$.

1.24 Let a, b be real numbers such that $|a|, |b| > k > 0$. Show that

$$\left| \frac{1}{a} - \frac{1}{b} \right| \leq \frac{|a - b|}{k^2}.$$

2
Sequences and Series

2.1 Sequences

The word "sequence" is used in mathematics with much the same meaning as in ordinary life, except that we always mean an infinite sequence. Thus

$$(2, 4, 6, 8, \ldots) \quad \text{and} \quad \left(\frac{1}{2}, \frac{2}{3}, \frac{3}{4}, \frac{4}{5}, \ldots \right)$$

are sequences. We often write the first of these as $(2n)_{n \in \mathbb{N}}$, or just as $(2n)$; and in the same way we can write the second sequence as $\left(n/(n+1) \right)_{n \in \mathbb{N}}$ or just as $\left(n/(n+1) \right)$.

If we want to talk about about sequences in general we use a notation like (a_n); thus $a_n = 2n$ in our first example, and $a_n = n/(n+1)$ in our second example. It is important to have this notation for more complicated sequences: for example, we can be quite precise about the sequence

$$\left(0, \frac{5}{4}, \frac{26}{27}, \frac{17}{16}, \frac{124}{125}, \frac{37}{36}, \ldots \right) \tag{2.1}$$

by describing it as (a_n), where

$$a_n = \begin{cases} (n^3 - 1)/n^3 & \text{if } n \text{ is odd} \\ (n^2 + 1)/n^2 & \text{if } n \text{ is even.} \end{cases}$$

How does a sequence differ from a set? We can give a precise answer to that question once we have discussed the general notion of a function in Chapter 3. For the moment, we can capture the essence by noting that *repetitions* and *order* are of no relevance in a set: the set $\{1, 3, 1, 3, \ldots\}$ is simply the set $\{1, 3\}$, or indeed the set $\{3, 1\}$. By contrast, the sequence $(1, 3, 1, 3, \ldots)$ is distinct from the sequence $(3, 1, 3, 1, \ldots)$. Two sequences (a_n) and (b_n) are equal if and only if $a_n = b_n$ for *all* n.

In many natural sequential processes one is interested in the long-term outcome, if this turns out to be a stable state. In mathematical terms this translates into an interest in the **limit** of the sequence. Intuitively we can see that the sequence $(n/(n + 1))$ has limit equal to 1, since the numbers

$$\frac{1}{2}, \frac{2}{3}, \frac{3}{4}, \frac{4}{5}, \ldots$$

get closer and closer to 1 as we go on. But this is arguably less clear (though still true) for the sequence (2.1), where some of the numbers are greater than 1 and others are less than 1, and where the difference between a_n and 1 does not steadily decrease.

What we need to say is that a sequence (a_n) has a limit α if a_n **can be made as close as we like to** α **by taking** n **large enough**. To put it more precisely, we say: **a sequence** (a_n) **has a limit** α **if, for every** $\epsilon > 0$ **there exists a natural number** N **such that** $|a_n - \alpha| < \epsilon$ **whenever** $n > N$.

This is not an easy definition to take in, which is not surprising, since it took about two centuries for the intuitive notion of limit to develop into the formal definition we have given. Think of ϵ as "small", though for very good reasons that has not been said.[1] After all, what is small depends on context: the diameter of the earth's orbit is large compared to the width of this page, but is small compared with the distance to the next galaxy. When we say that (a_n) has limit α we are saying, in effect, that no matter how small the positive number ϵ may be, we can arrange for $|a_n - \alpha|$ to be less than ϵ for all n beyond a certain N. This N will of course depend upon ϵ, and will normally become larger if we choose a smaller ϵ.

In this spirit, let us look again at the sequence

$$\left(\frac{1}{2}, \frac{2}{3}, \frac{3}{4}, \ldots \right),$$

where $a_n = n/(n + 1)$. This has limit 1. Notice that

$$|a_n - 1| = \left| \frac{n}{n + 1} - 1 \right| = \left| \frac{-1}{n + 1} \right| = \frac{1}{n + 1}.$$

Given ϵ, we have that

$$\frac{1}{n + 1} < \epsilon \text{ whenever } n > \frac{1}{\epsilon} - 1. \tag{2.2}$$

---

[1] The use of the Greek letter ϵ is completely standard in this context – to the extent that formal, rigorous analysis has sometimes been called "epsilonics".

So if we take N as any positive integer not less than $(1/\epsilon) - 1$, we can say that

$$|a_n - \alpha| < \epsilon \text{ whenever } n > N.$$

It is evident that N depends on ϵ, and it is instructive to make a table:

ϵ	0.01	0.001	0.0001	...
N	99	999	9999	...

Again, let us look at the sequence (2.1). Here we see that

$$|a_n - 1| = \begin{cases} 1/n^3 & \text{if } n \text{ is odd} \\ 1/n^2 & \text{if } n \text{ is even.} \end{cases}$$

Given $\epsilon > 0$, let N be any (positive) integer such that

$$N \geq 1/\sqrt{\epsilon}; \tag{2.3}$$

then, for all odd $n > N$,

$$|a_n - 1| = \frac{1}{n^3} < \frac{1}{n^2} < \frac{1}{N^2} \leq \epsilon,$$

and, for all even $n > N$,

$$|a_n - 1| = \frac{1}{n^2} < \frac{1}{N^2} \leq \epsilon.$$

We have shown that for every $\epsilon > 0$ there exists N such that $|a_n - 1| < \epsilon$ for every $n > N$. Thus $(a_n) \to 1$. Here the table showing the dependence of N on ϵ is

ϵ	0.01	0.001	0.0001	...
N	10	32	100	...

One useful way of looking at the notion of limit is to imagine an argument in which your opponent hands you an ϵ and challenges you to find an appropriate N. In the cases we have considered, you can always win the argument, because you have the secret formulae (2.2) and (2.3), and so you can establish that the limit is indeed 1. By contrast, consider an (admittedly rather contrived) sequence

$$(0.9998, 0.9999, 0.9998, 0.9999, \ldots), \tag{2.4}$$

and let us examine the claim that the limit is 1. If your opponent chooses $\epsilon = 0.01$ then you can respond with confidence that

$$|a_n - 1| < 0.01 \text{ for all } n > 1,$$

and even if your opponent chooses $\epsilon = 0.001$ you can win the argument, since

$$|a_n - 1| < 0.001 \text{ for all } n > 1.$$

But if your opponent chooses $\epsilon = 0.0001$, then there is no natural number N for which $|a_n - 1| < 0.0001$ for all $n > N$, and you are forced to conclude that the limit is not 1. For the limit to be 1 you have to win the argument *every* time, no matter what ϵ your opponent chooses. The sequence (2.4) in fact has no limit, for there is no number that is simultaneously arbitrarily close to 0.9998 and 0.9999.

You may find it helpful to visualise a sequence in a graphical way:

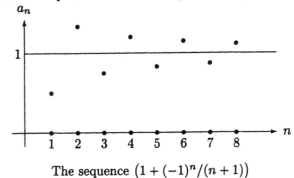

The sequence $\left(1 + (-1)^n/(n+1)\right)$

If a sequence (a_n) has a limit α we say that it is **convergent** (to α), and we write

$$\lim_{n \to \infty} a_n = \alpha, \text{ or, more simply, } (a_n) \to \alpha.$$

Otherwise, the sequence is said to be **divergent**. Examples of divergent sequences are

$$(2, 4, 6, 8, \ldots), \quad (1, -1, 1, -1, \ldots), \quad (-1, 2, -3, 4, -5, \ldots).$$

Theorem 2.1

Let (a_n) be a convergent sequence, with limit α. Then $(|a_n|)$ is convergent, with limit $|\alpha|$.

Proof

Our assumption is that for all $\epsilon > 0$ there exists N such that $|a_n - \alpha| < \epsilon$ for all $n > N$. Now by Theorem 1.9 we have that $||a_n| - |\alpha|| \leq |a_n - \alpha|$, and so it follows that

$$||a_n| - |\alpha|| < \epsilon$$

for all $n > N$. Thus $(|a_n|) \to |\alpha|$, as required. $\qquad\qquad\square$

Remark 2.2

The converse of this theorem is untrue in general. For example, if $a_n = (-1)^{n+1}$, so that we have the sequence $(1, -1, 1, -1, \ldots)$, then (a_n) is not convergent, but $(|a_n|)$, being the constant sequence $(1, 1, 1, \ldots)$, clearly has limit 1.

We do, however, have the following converse, concerning **null** sequences (sequences with limit 0):

Theorem 2.3

Let (a_n) be a sequence of real numbers. Then $(a_n) \to 0$ if and only if $(|a_n|) \to 0$.

Proof

By virtue of Theorem 2.1 we need only prove "if". So suppose that $(|a_n|) \to 0$. Then for every $\epsilon > 0$ there exists N such that $||a_n| - 0| < \epsilon$ for all $n > N$. Now

$$||a_n| - 0| = ||a_n|| = |a_n| = |a_n - 0|,$$

and so we deduce that $|a_n - 0| < \epsilon$ for all $n > N$. Thus $(a_n) \to 0$, as required. $\qquad \square$

One possibility for a divergent sequence (a_n) is that it **tends to infinity**. Approaching this question in the same spirit as for the statement $(a_n) \to \alpha$, we would wish to say that a_n can be made arbitrarily large by taking n large enough. More precisely, we say[2] that $(a_n) \to \infty$ **if for all $K > 0$ there exists a natural number N such that $a_n > K$ for all $n > N$.**

Theorem 2.4

Let (a_n) be a sequence with the property that $a_n > 0$ for all $n \geq 1$. Then $(a_n) \to \infty$ if and only if $(1/a_n) \to 0$.

Proof

Suppose first that $(a_n) \to \infty$. Let $\epsilon > 0$. Then $1/\epsilon > 0$, and by our assumption (if we take $K = 1/\epsilon$) there exists N such that $a_n > 1/\epsilon$ for all $n > N$. Since both a_n and $1/\epsilon$ are positive, we deduce that

$$\left| \frac{1}{a_n} - 0 \right| = \left| \frac{1}{a_n} \right| = \frac{1}{a_n} < \epsilon$$

---

[2] We write $(a_n) \to \infty$, but avoid writing $\lim_{n \to \infty} a_n = \infty$ lest we seem to be claiming that there is a real number called ∞. There is no such number.

for all $n > N$. Hence $(1/a_n) \to 0$.

Conversely, suppose that $(1/a_n) \to 0$. Let K be a positive number. Then $1/K$ is also positive, and so there exists N such that $|1/a_n| < 1/K$ for all $n > N$. Hence, since $|1/a_n| = 1/a_n$, we deduce that $a_n > K$ for all $n > N$, and so $(a_n) \to \infty$. $\square$

Remark 2.5

The stipulation in the above theorem that $a_n > 0$ is necessary. If, for example, $a_n = (-1)^{n+1}n$, so that we have the sequence $(1, -2, 3, -4, \ldots)$, then $(1/a_n) \to 0$, but we cannot say that $(a_n) \to \infty$.

If (a_n) is a sequence, we say that $(a_n) \to -\infty$ if $(-a_n) \to \infty$.

One important sequence that we shall have many occasions to use is the sequence $(b^n)_{n \in \mathbb{N}}$, where b is a fixed number. If $b = 1$, then (b^n) is the constant sequence $(1, 1, 1, \ldots)$, and this has limit 1. Similarly, if $b = 0$ it is immediate that $(b^n) \to 0$.

Suppose now that $b > 1$. We may write $b = 1 + c$, where $c > 0$, and so, by the binomial theorem (Theorem 1.7),

$$b^n = 1 + nc + \binom{n}{2}c^2 + \binom{n}{3}c^2 + \cdots + c^n .$$

All the terms in the binomial expansion are positive, and so certainly $b^n \geq 1 + nc$. Let K be a positive real number, and let N be any integer such that $N > (K-1)/c$. Then for all $n > N$,

$$b^n \geq 1 + nc > 1 + (K-1) = K ,$$

and we deduce that $(b^n) \to \infty$.

Suppose now that $0 < b < 1$. Then $1/b > 1$ and so $(1/b^n) \to \infty$. Hence, by Theorem 2.4, $(b^n) \to 0$.

Suppose next that $-1 < b < 0$. Then $0 < |b| < 1$ and so the sequence $(|b^n|) = (|b|^n)$ has limit 0. By Theorem 2.3, we deduce that $(b^n) \to 0$.

Suppose next that $b < -1$. Then $|b| > 1$. If (b^n) were convergent, then by Theorem 2.1 we would deduce that $(|b|^n)$ was convergent. Since this is not the case, we must have that (b^n) is divergent. Here we cannot say that $(b^n) \to \infty$: if, for example, $b = -2$, we obtain the sequence

$$(-2, 4, -8, 16, -32, \ldots)$$

which contains both positive and negative terms.

Noting finally that for $b = -1$ we obtain the divergent sequence

$$(-1, 1, -1, 1, \ldots),$$

we can summarise our main results in a theorem as follows:

Theorem 2.6

Let b be a real number. Then:

(i) the sequence (b^n) converges if and only if $b \in (-1, 1]$;

(ii) $(b^n) \to 0$ if $b \in (-1, 1)$.

EXERCISES

2.1 The sequence $(1/\sqrt{n})$ has limit 0. For each of $\epsilon = 0.01, 0.001, 0.0001$, determine an integer N with the property that $|(1/\sqrt{n}) - 0| < \epsilon$ for all $n > N$.

2.2 Show that the sequence $(1/n^k)_{n \in \mathbb{N}}$ is convergent if and only if $k \geq 0$, and that the limit is 0 for all $k > 0$.

2.3 Determine the least value of N such that $n/(n^2 + 1) < 0.0001$ for all $n \geq N$.

2.4 Determine the least value of N such that $n^2 + 2n \geq 9999$ for all $n > N$.

2.5 Give a formal definition of the statement $(a_n) \to -\infty$.

2.6 Let $a_1 = 0$, $a_2 = 3$, and, for all $n \geq 3$ let

$$a_n = \frac{1}{2}(a_{n-1} + a_{n-2}).$$

(The sequence (a_n) is said to be defined **recursively**.) By induction on n, show that, for all $n \geq 2$,

$$a_n = 2 + 4\left(-\frac{1}{2}\right)^n,$$

and deduce that $(a_n) \to 2$.

2.2 Sums, Products and Quotients

The process of finding an N for each ϵ can be quite complicated, and it is important to prove some theorems that will enable us to avoid the process as much as possible.

Before we introduce the main theorem (Theorem 2.8) of this section we
have to make some further definitions. A sequence (a_n) is said to be **bounded
above** if there exists a real number K such that $a_n \leq K$ for all $n \geq 1$, and to
be **bounded below** if there exists a real number L such that $a_n \geq L$ for all
$n \geq 1$. The sequence (a_n) is said to be **bounded** if there exists a real number
M such that $|a_n| \leq M$ for all $n \geq 1$. It is not hard to show (see Exercise 2.8)
that (a_n) is bounded if and only if it is bounded above and below.

A bounded sequence need not be convergent – consider, for example, the
sequence $((-1)^{n+1})$. But the converse is true:

Theorem 2.7

Every convergent sequence is bounded.

Proof

Let (a_n) be a sequence with limit α. Then, by Theorem 2.1, $(|a_n|)$ has limit
$|\alpha|$. Quite arbitrarily, let us choose $\epsilon = 1$, and then use the definition of a limit
to conclude that there exists N such that $||a_n| - |\alpha|| < 1$ for all $n > N$. That
is, for all $n > N$,
$$|\alpha| - 1 < |a_n| < |\alpha| + 1\,.$$
Hence, for all $n \geq 1$,
$$|a_n| \leq \max\left\{|a_1|, |a_2|, \ldots, |a_N|, |\alpha| + 1\right\},$$
and so (a_n) is bounded. □

The following result is a very powerful tool in determining limits:

Theorem 2.8

Let (a_n), (b_n) be sequences such that $(a_n) \to \alpha$, $(b_n) \to \beta$. Then:

(i) $(-a_n) \to -\alpha$;

(ii) $(a_n + b_n) \to \alpha + \beta$;

(iii) $(a_n - b_n) \to \alpha - \beta$;

(iv) $(a_n b_n) \to \alpha\beta$;

(v) $(ka_n) \to k\alpha$ for every constant k;

(vi) $(1/b_n) \to 1/\beta$, provided b_n is non-zero for all n, and provided $\beta \neq 0$;

(vii) $(a_n/b_n) \to \alpha/\beta$, provided b_n is non-zero for all n, and provided $\beta \neq 0$.

Proof

(i) For every $\epsilon > 0$ there exists N such that $|a_n - \alpha| < \epsilon$ for all $n > N$. Since

$$|-a_n - (-\alpha)| = |a_n - \alpha|,$$

it follows that $|-a_n - (-\alpha)| < \epsilon$ for all $n > N$, and so $(-a_n) \to -\alpha$.

(ii) Let $\epsilon > 0$. Since $(a_n) \to \alpha$, we can make $|a_n - \alpha|$ as small as we desire by taking n sufficiently large, and for technical reasons that will be apparent in a moment, we choose to make it smaller than $\epsilon/2$. That is, there exists N_1 such that $|a_n - \alpha| < \epsilon/2$ for all $n > N_1$. Equally, since $(b_n) \to \beta$, there exists N_2 such that $|b_n - \beta| < \epsilon/2$ for all $n > N_2$. Hence, for all $n > N = \max\{N_1, N_2\}$,

$$|(a_n + b_n) - (\alpha + \beta)| = |(a_n - \alpha) + (b_n - \beta)| \leq |(a_n - \alpha)| + |(b_n - \beta)| < \frac{\epsilon}{2} + \frac{\epsilon}{2} = \epsilon,$$

and so $(a_n + b_n) \to \alpha + \beta$.

(iii) This follows from (i) and (ii), since we may write $(a_n - b_n)$ as $(a_n + (-b_n))$.

(iv) By Theorem 2.7, we may assume that there are positive numbers A and B such that $|a_n| \leq A$ and $|b_n| \leq B$ for all $n \geq 1$. By Theorem 2.1 and Exercise 2.7 below, we may deduce that $|\beta| \leq B$ also. Let $\epsilon > 0$. Then there exists N_3 such that $|a_n - \alpha| < \epsilon/2B$ for all $n > N_3$, and there exists N_4 such that $|b_n - \beta| < \epsilon/2A$ for all $n > N_4$. By a standard and very important algebraic trick it then follows that, for all $n > N = \max\{N_3, N_4\}$,

$$\begin{aligned} |a_n b_n - \alpha\beta| &= |a_n(b_n - \beta) + \beta(a_n - \alpha)| \\ &\leq |a_n||b_n - \beta| + |\beta||a_n - \alpha| \\ &\leq A(\epsilon/2A) + B(\epsilon/2B) = \epsilon. \end{aligned}$$

Thus $(a_n b_n) \to \alpha\beta$, as required.

(v) This follows immediately from (iv) if we take (b_n) as the constant sequence $(k, k, k, \ldots)$.

(vi) Since $\beta \neq 0$ we may take $\epsilon = |\beta|/2$ and assert that there exists N_5 such that $|b_n - \beta| < |\beta|/2$ for all $n > N_5$. Hence

$$|\beta| - |b_n| \leq ||b_n| - |\beta|| \leq |b_n - \beta| < |\beta|/2,$$

and so $|b_n| > |\beta|/2$. It follows that $1/|b_n| < 2/|\beta|$. Now let $\epsilon > 0$ be given. Then there exists N_6 such that $|b_n - \beta| < \epsilon|\beta|^2/2$ for all $n > N_6$. It follows that, for all $n > \max\{N_5, N_6\}$,

$$\left|\frac{1}{\beta} - \frac{1}{b_n}\right| = \frac{|b_n - \beta|}{|b_n \beta|} < \frac{2|b_n - \beta|}{|\beta^2|} < \epsilon,$$

and so $(1/b_n) \to 1/\beta$.

(vii) This follows from (iv) and (vi), since we may consider (a_n/b_n) as $(a_n(1/b_n))$.   □

Theorem 2.8 has many applications. If, for example, we wish to find $\lim_{n\to\infty} a_n$, where

$$a_n = \frac{3n^3 - 7n + 2}{4n^3 + 8n^2},$$

we can divide the numerator and the denominator of the fraction by n^3, and write

$$a_n = \frac{3 - (7/n^2) + (2/n^3)}{4 + (8/n)}.$$

Now, it is clear that the constant sequences (3) and (4) have limits 3 and 4 respectively. Also, the sequences $(1/n^2)$, $(1/n^3)$ and $(1/n)$ all have limit 0. We deduce from Theorem 2.8 that

$$\lim_{n\to\infty} a_n = \frac{3 - 0 + 0}{4 + 0} = \frac{3}{4}.$$

EXERCISES

2.7 Let (b_n) be a sequence with limit β. Show that if B is an upper bound for (b_n), then $\beta \leq B$.

2.8 Show that a sequence (a_n) is bounded if and only if it is bounded above and below.

2.9 Let (a_n) be a sequence with limit α, and define $b_n = a_{n+1}$ $(n = 1, 2, \ldots)$. Show that $(b_n) \to \alpha$.

2.10 More generally, let (b_n) be a sequence obtained from (a_n) by deleting terms so that infinitely many remain. Show that if $(a_n) \to \alpha$ then $(b_n) \to \alpha$ also.

2.11 Show that, if $a_n \geq 0$ for all $n \geq 1$ and if $(a_n) \to L$, then $L \geq 0$.

2.12 Let (a_n) be a sequence of positive terms, and suppose that $(a_n) \to L$. Show that $(\sqrt{a_n}) \to \sqrt{L}$.

2.13 Let (a_n), (x_n) and (b_n) be sequences with limits α, ξ, β, respectively, and suppose that, for all $n \geq 1$,

$$a_n \leq x_n \leq b_n.$$

Show that $\alpha \leq \xi \leq \beta$. [This is sometimes called the **sandwich principle**. It can be useful if (a_n) and (b_n) are "known" sequences and (x_n) is unknown. It is especially useful when $\alpha = \beta$, for in this case we conclude that $\xi = \alpha = \beta$.]

2.14 Using the identities

$$\max\{x, y\} = \frac{1}{2}(x + y + |x - y|), \quad \min\{x, y\} = \frac{1}{2}(x + y - |x - y|),$$

show that if (a_n), (b_n) are sequences such that $(a_n) \to \alpha$, $(b_n) \to \beta$, then

$$\left(\max\{a_n, b_n\}\right) \to \max\{\alpha, \beta\}, \quad \left(\min\{a_n, b_n\}\right) \to \min\{\alpha, \beta\}.$$

Give an example of divergent sequences (a_n), (b_n) for which the sequences $\left(\max\{a_n, b_n\}\right)$ and $\left(\min\{a_n, b_n\}\right)$ both converge. Show, however, that if (a_n), $\left(\max\{a_n, b_n\}\right)$ and $\left(\min\{a_n, b_n\}\right)$ are all convergent, then (b_n) is also convergent.

2.15 Consider the sequence $(a^{1/n})$, where $a > 1$. Use the binomial theorem to show that $a > 1 + nh_n$, where $h_n = a^{1/n} - 1$. Deduce that $(a^{1/n}) \to 1$. Show that this holds also for $0 < a \leq 1$.

2.16 Determine $\lim_{n \to \infty} (2^n + 3^n)^{1/n}$.

2.17 Let (a_n) be a sequence such that $(a_n^3) \to \alpha^3$, where $\alpha \neq 0$. Show that

$$|a_n - \alpha| \leq \frac{|a_n^3 - \alpha^3|}{(3/4)\alpha^2},$$

and deduce that $(a_n) \to \alpha$.

a) Does the result hold when $\alpha = 0$?

b) Does the result hold if the cube is replaced by the square?

2.3 Monotonic Sequences

It is time to develop some theory. We say that a sequence (a_n) is **monotonic increasing** if $a_{n+1} \geq a_n$ for all $n \geq 1$, and **ultimately monotonic increasing** if there exists $N \geq 1$ such that $a_{n+1} \geq a_n$ for all $n \geq N$. The definitions of **monotonic decreasing** and **ultimately monotonic decreasing** are obvious, and we say that (a_n) is **[ultimately] monotonic** if it is either [ultimately] monotonic increasing or [ultimately] monotonic decreasing. For example, the sequence $(\frac{n-1}{n})$ is monotonic increasing, and the sequence $(\frac{n+1}{n})$ is monotonic decreasing. The sequence

$$\left(1 + \frac{(-1)^{n+1}}{n}\right)_{n \in \mathbf{N}} = \left(2, \frac{1}{2}, \frac{4}{3}, \frac{3}{4}, \frac{6}{5}, \frac{5}{6}, \cdots\right)$$

is not monotonic.

The following result is of enormous importance in the development of the theory:

Theorem 2.9

Let (a_n) be an ultimately monotonic increasing sequence bounded above by K. Then (a_n) is convergent.

Proof

Suppose that $a_{n+1} \geq a_n$ for all $n \geq N_1$. Since the set $\{a_n : n \geq N_1\}$ is bounded above by K, it has, by the completeness axiom, a supremum L. We show that $(a_n) \to L$. Let $\epsilon > 0$ be given. Then, since L is the *least* upper bound, $L - \epsilon$ is not an upper bound for $\{a_n : n \geq N_1\}$. Hence there exists a natural number $N_2 \geq N_1$ such that $L - \epsilon < a_{N_2} \leq L$. Since (a_n) is ultimately increasing, it follows that $L - \epsilon < a_n \leq L$ for all $n > N_2$. Thus $|a_n - L| < \epsilon$ for all $n > N_2$, and so $(a_n) \to L$, as required. $\square$

A very similar argument shows:

Theorem 2.10

Let (a_n) be an ultimately monotonic decreasing sequence bounded below by K. Then (a_n) is convergent.

These two theorems are the first "existence theorems" in our account of sequences, in that they assert that a limit exists without providing a direct way of finding the limit.

Example 2.11

Consider a sequence (a_n) defined by

$$a_1 = 2, \quad a_{n+1} = \frac{1}{2}\left(a_n + \frac{2}{a_n}\right) \quad (n \geq 1).$$

Show that $(a_n) \to \sqrt{2}$.

Solution

Since it is clear that a_n is positive for all n, we see that the sequence (a_n) is bounded below by 0. It is, however, useful to establish that there is a better lower bound than this. It is clear that $a_1^2 > 2$, and we show by induction that

$$a_n^2 > 2 \text{ for all } n \geq 1.\tag{2.5}$$

For suppose that $a_k^2 > 2$. Then

$$a_{n+1}^2 - 2 = \frac{1}{4}\left(a_n^2 + 4 + \frac{4}{a_n^2} - 8\right)$$

$$= \frac{1}{4}\left(a_n^2 - 4 + \frac{4}{a_n^2}\right) = \frac{(a_n^2 - 2)^2}{4a_k^2} > 0.$$

Thus, taking (positive) square roots, we see that $a_n > \sqrt{2}$ for all $n \geq 1$. Next, we show that $a_{n+1} < a_n$ for all $n \geq 1$. To see this, observe that, by (2.5),

$$a_n - a_{n+1} = a_n - \frac{1}{2}\left(a_n + \frac{2}{a_n}\right) = \frac{1}{2}\left(a_n - \frac{2}{a_n}\right) = \frac{a_n^2 - 2}{2a_n} > 0.$$

Since (a_n) is a decreasing sequence, bounded below by $\sqrt{2}$, we deduce that it has a limit α. As yet we do not know what α is. However, if we define

$$b_n = \frac{1}{2}\left(a_n + \frac{2}{a_n}\right),$$

we know from Theorem 2.8 that

$$(b_n) \to \frac{1}{2}\left(\alpha + \frac{2}{\alpha}\right).$$

On the other hand, it is clear that the sequence (a_{n+1}) also has limit α (see Exercise 2.10), and so we deduce that

$$\alpha = \frac{1}{2}\left(\alpha + \frac{2}{\alpha}\right).$$

From this we easily deduce that $\alpha^2 = 2$, and hence, since α is certainly positive, that $\alpha = \sqrt{2}$. $\qquad\square$

Remark 2.12

This is actually quite a practical way of computing $\sqrt{2}$, since the convergence to $\sqrt{2}$ is very rapid. An easy calculation shows that $a_4 = 577/408$, whose square is approximately 2.000006.

We turn now to an example of great importance in the development of the theory.

Example 2.13

Show that the sequence (a_n) given by

$$a_n = \left(1 + \frac{1}{n}\right)^n \quad (n = 1, 2, \ldots)$$

is convergent.

Solution

A preliminary examination suggests that (a_n) might be a monotonic increasing sequence:

$$a_1 = 2, \ a_2 = 9/4 = 2.25, \ a_3 = 64/27 \approx 2.37, \ a_4 = 625/256 \approx 2.44.$$

(Read the symbol $\approx$ as "is approximately equal to".) To see that this is actually the case, we use the binomial theorem to obtain

$$a_n = \sum_{r=0}^{n} \binom{n}{r} \frac{1}{n^r}, \tag{2.6}$$

$$a_{n+1} = \sum_{r=0}^{n+1} \binom{n+1}{r} \frac{1}{(n+1)^r}.$$

Hence

$$a_{n+1} - a_n = \sum_{r=0}^{n} \left[\binom{n+1}{r} \frac{1}{(n+1)^r} - \binom{n}{r} \frac{1}{n^r} \right] + \frac{1}{(n+1)^{n+1}}. \tag{2.7}$$

Now,

$$\binom{n}{r} \frac{1}{n^r} = \frac{n(n-1)\ldots(n-r+1)}{r!\,n^r}$$

$$= \frac{1}{r!} \left(1 - \frac{1}{n}\right) \left(1 - \frac{2}{n}\right) \ldots \left(1 - \frac{r-1}{n}\right), \tag{2.8}$$

and similarly

$$\binom{n+1}{r} \frac{1}{(n+1)^r} = \frac{1}{r!} \left(1 - \frac{1}{n+1}\right) \left(1 - \frac{2}{n+1}\right) \ldots \left(1 - \frac{r-1}{n+1}\right). \tag{2.9}$$

Each factor $1 - (i/(n+1))$ in (2.9) is greater than the corresponding factor $1 - (i/n)$ in (2.8). Hence from (2.7) we may deduce that $a_{n+1} - a_n > 0$.

Next, we show that (a_n) is bounded above. In (2.8), each of the factors $1 - (i/n)$ is less than 1, and so from (2.6) we deduce that

$$a_n < \sum_{r=0}^{n} \frac{1}{r!}.$$

Since it is clear that, for all $r \geq 2$,

$$r! = 2.3.\ldots.r \geq 2^{r-1},$$

it follows that

$$a_n < 1 + \sum_{r=0}^{n} 2^{r-1} < 1 + \frac{1}{1 - (1/2)} = 3.$$

We deduce that (a_n) converges to a limit,[3] usually called e, where $2 < e < 3$. $\square$

The final worked example of this section is somewhat easier:

Example 2.14

Determine $\lim_{n \to \infty} a_n$, where

$$a_1 = 1, \quad a_n = \sqrt{1 + 2a_{n-1}} \quad (n \geq 2).$$

Solution

If a limit α exists, then it is non-negative, and, by letting $n \to \infty$ in the formula $a_n = \sqrt{1 + 2a_{n-1}}$, we see that $\alpha^2 = 1 + 2\alpha$. Hence $\alpha = 1 + \sqrt{2}$, the positive root of the equation $\alpha^2 - 2\alpha - 1 = 0$.

So does the limit exist? Notice that, for positive x,

$$x^2 - 2x - 1 < 0 \text{ if and only if } x < \alpha. \tag{2.10}$$

We show by induction that $a_n < \alpha$ for all n. Certainly $a_1 < \alpha$, and, if we suppose inductively that $a_{n-1} < \alpha$, we can deduce that

$$\alpha^2 - a_n^2 = \alpha^2 - 2a_{n-1} - 1 > \alpha^2 - 2\alpha - 1 = 0.$$

Thus $a_n < \alpha$, since both a_n and α are positive. Next, by (2.10),

$$a_{n-1}^2 - a_n^2 = a_{n-1}^2 - 2a_{n-1} - 1 < 0,$$

and so $a_n > a_{n-1}$ for all $n \geq 2$.

We now know that (a_n) is monotonic increasing, bounded above by α. Hence a limit exists, and we have seen that the limit must be $\alpha = 1 + \sqrt{2}$. $\square$

---

[3] We shall come across this number again!

EXERCISES

2.18 Consider the sequence (a_n), where

$$a_1 = 0, \qquad a_{n+1} = \frac{3a_n + 1}{a_n + 3} \quad (n \geq 1).$$

a) Show by induction that $0 \leq a_n < 1$ for all n.

b) Show that (a_n) is monotonic increasing.

c) Find $\lim_{n \to \infty} a_n$.

2.19 Consider the sequence (a_n), where

$$a_1 = 2, \qquad a_{n+1} = \frac{2a_n + 3}{a_n + 2} \quad (n \geq 1).$$

a) Show by induction that $a_n^2 > 3$ for all n.

b) Show that (a_n) is monotonic decreasing.

c) Find $\lim_{n \to \infty} a_n$.

2.20 Determine the limit of (a_n), where

$$a_1 = 2, \qquad a_{n+1} = \sqrt{2 + 2a_n} \quad (n \geq 1).$$

2.21 Let (a_n), (b_n) be sequences of positive numbers, such that, for all $n \geq 1$,

$$a_{n+1} = \frac{1}{2}(a_n + b_n), \quad b_{n+1} = \sqrt{a_n b_n}.$$

a) Show that, from $n = 2$ onwards, (a_n) is monotonic decreasing and (b_n) is monotonic increasing.

b) Deduce that (a_n) and (b_n) have the same limit.

2.4 Cauchy Sequences

Of course not every sequence has the friendly property of being monotonic. We now investigate a more general technique that can establish the existence of a limit even when the sequence is not monotonic. The origin of the idea is simple enough: if a sequence (a_n) has the property that its members get closer and closer to some limit α as n increases, then they also get closer and closer to each other. To put this more formally, we say that (a_n) is a **Cauchy sequence** if, for every $\epsilon > 0$ there exists $N \geq 1$ such that $|a_m - a_n| < \epsilon$ for all $m, n > N$. Then the following result is easily proved:

Theorem 2.15

Every convergent sequence is a Cauchy sequence.

Proof

Let (a_n) be a sequence with limit α, and let $\epsilon > 0$ be given. Then there exists $N \geq 1$ such that $|a_n - \alpha| < \epsilon/2$ for all $n > N$. It follows that, for all $m, n > N$,

$$|a_m - a_n| = |(a_m - \alpha) + (\alpha - a_n)| \leq |a_m - \alpha| + |a_n - \alpha| < \frac{\epsilon}{2} + \frac{\epsilon}{2} = \epsilon,$$

and so (a_n) is a Cauchy sequence. $\qquad\square$

The converse of this result is harder to prove, but is of great importance:

Theorem 2.16

Every Cauchy sequence is convergent.

Proof

Let (a_n) be a Cauchy sequence. We show first that (a_n) is bounded. In the definition of a Cauchy sequence we are certainly entitled to take $\epsilon = 1$ and to conclude that there exists $N \geq 1$ with the property that $|a_m - a_n| < 1$ for all $m, n > N$. In particular, taking $n = N + 1$, we have that $|a_m - a_{N+1}| < 1$ for all $m > N$; that is,

$$a_{N+1} - 1 < a_m < a_{N+1} + 1$$

for all $m > N$. We deduce that, for all $m \geq 1$,

$$a_m \leq \max\{a_1, \ldots, a_N, a_{N+1} + 1\} = C,$$
$$a_m \geq \min\{a_1, \ldots, a_N, a_{N+1} - 1\} = D. \qquad (2.11)$$

Hence the sequence (a_n) is bounded.

To show that (a_n) is convergent, notice first that each of the sets

$$A_m = \{a_m, a_{m+1}, \ldots\} \quad (m = 1, 2, 3, \ldots)$$

is bounded. Moreover,

$$A_1 \supseteq A_2 \supseteq A_3 \supseteq \cdots.$$

If we now define $K_m = \sup A_m$ for $m = 1, 2, 3, \ldots$, we deduce that

$$K_1 \geq K_2 \geq K_3 \geq \cdots,$$

and so $(K_m)_{m \in \mathbb{N}}$ is a monotonic decreasing sequence. It is also bounded below, since (with D defined as in (2.11))

$$K_m = \sup\{a_n : n \geq m\} \geq a_m \geq D.$$

By Theorem 2.10 we deduce that (K_m) converges to a limit L.

We show that $(a_n) \to L$. Let $\epsilon > 0$ be given. Since (a_n) is by assumption a Cauchy sequence, we may choose $N_1 \geq 1$ so that $|a_m - a_n| < \epsilon/3$ for all $m, n > N_1$. We may also choose $N_2 \geq 1$ so that $|K_n - L| < \epsilon/3$ for all $n > N_2$. Let $n > \max\{N_1, N_2\}$. Then there exists a_m in $A_n = \{a_k : k \geq n\}$ for which $a_m > K_n - \epsilon/3$, since otherwise we would have $a_m \leq K_n - \epsilon/3$ for all $m \geq n$, and we would have an upper bound for A_n smaller than K_n. Thus

$$|K_n - a_m| = K_n - a_m < \epsilon/3.$$

It now follows that, for all $n > \max\{N_1, N_2\}$, and with m chosen in the way we have indicated,

$$\begin{aligned}
|a_n - L| &= |(a_n - a_m) + (a_m - K_n) + (K_n - L)| \\
&\leq |a_m - a_n| + |K_n - a_m| + |K_n - L| \\
&< \frac{\epsilon}{3} + \frac{\epsilon}{3} + \frac{\epsilon}{3} = \epsilon.
\end{aligned}$$

Hence (a_n) has limit L. $\square$

We finish this section by examining two sequences that are not monotonic.

Example 2.17

Let

$$q_n = \begin{cases} 1/2 & \text{if } n \text{ is prime} \\ -1 & \text{otherwise,} \end{cases}$$

and let (a_n), be defined recursively by

$$a_1 = 1, \quad a_{n+1} = a_n + 2^{-n} q(n) \ (n \geq 1).$$

Show that (a_n) is a Cauchy sequence.

Solution

Let $\epsilon > 0$ be given. Choose N so that $1/2^N < \epsilon$, and let $m > n > N$. Then

$$\begin{aligned}
|a_m - a_n| &= |(a_{n+1} - a_n) + (a_{n+2} - a_{n+1}) + \cdots + (a_m - a_{m-1})| \\
&\leq |a_{n+1} - a_n| + |a_{n+2} - a_{n+1}| + \cdots + |a_m - a_{m-1}| \\
&\leq 2^{-n} + 2^{-(n+1)} + \cdots + 2^{-(m-1)} \quad \text{(since } |q_n| \leq 1 \text{ for all } n)
\end{aligned}$$

$$\leq \frac{2^{-n}}{1 - (1/2)} = \frac{1}{2^{n-1}} \quad \text{(by Exercise 1.19)}$$

$$\leq \frac{1}{2^N} < \epsilon.$$

The sequence is a Cauchy sequence, and so has a limit. Here we are not able easily to determine the exact value of the limit. $\qquad\square$

One of the most important sequences in mathematics is the **Fibonacci**[4] sequence

$$(1, 1, 2, 3, 5, 8, 13, 21, \ldots),$$

defined as (f_n), where

$$f_1 = f_2 = 1, \quad f_n = f_{n-1} + f_{n-2}, \ (n \geq 3).$$

The sequence (f_n) is of course divergent, but if we define $a_n = f_n/f_{n+1}$ for all $n \geq 1$ we obtain a sequence

$$1, 0.5, 0.667, 0.6, 0.625, \ldots$$

that looks as if it might converge. It is clearly not monotonic, but, since (f_n) is an increasing sequence, it is clear that $a_n \leq 1$ for all n. Only slightly less clear is the observation that, for all $n \geq 1$,

$$a_n = \frac{f_n}{f_n + f_{n-1}} \geq \frac{f_n}{f_n + f_n} = \frac{1}{2}.$$

Next, notice that, for all $n \geq 1$,

$$1/a_{n+1} = \frac{f_{n+2}}{f_{n+1}} = \frac{f_{n+1} + f_n}{f_{n+1}} = 1 + a_n,$$

and so

$$a_{n+1} = \frac{1}{1 + a_n}. \tag{2.12}$$

It follows that

$$a_{n+1} - a_n = \frac{1}{1 + a_n} - \frac{1}{1 + a_{n-1}} = \frac{a_{n-1} - a_n}{(1 + a_n)(1 + a_{n-1})}.$$

Since, for all $n \geq 2$,

$$(1 + a_n)(1 + a_{n-1}) \geq (1 + \frac{1}{2})(1 + \frac{1}{2}) = \frac{9}{4} > 2,$$

---

[4] Leonardo of Pisa (c. 1170–1250) was also known as Fibonacci, the "son of Bonaccio".

we can assert that for all $n \geq 2$

$$|a_{n+1} - a_n| \leq \frac{1}{2}|a_n - a_{n-1}|$$

Replacing n by $n-1$ in this inequality, we obtain

$$|a_n - a_{n-1}| \leq \frac{1}{2}|a_{n-1} - a_{n-2}|$$

for all $n \geq 3$, and hence

$$|a_{n+1} - a_n| \leq \frac{1}{4}|a_{n-1} - a_{n-2}|\,.$$

More generally, by repeating the argument, we see that, for all $n \geq 2$,

$$|a_{n+1} - a_n| \leq \frac{1}{2^{n-3}}|a_3 - a_2| = \frac{1}{2^{n-3}}\,.$$

Suppose now that N is a natural number not less than 3, and let $m > n > N$. Then

$$|a_m - a_n| = |(a_m - a_{m-1}) + \cdots + (a_{n+2} - a_{n+1}) + (a_{n+1} - a_n)|$$
$$\leq |a_m - a_{m-1}| + \cdots + |a_{n+2} - a_{n+1}| + |a_{n+1} - a_n|.$$

For each k such that $n \leq k \leq m-1$, we know that $|a_{k+1} - a_k| \leq 1/2^{k-3}$. Hence

$$|a_m - a_n| \leq \left(\frac{1}{2^{n-3}} + \frac{1}{2^{n-2}} + \cdots + \frac{1}{2^{m-4}} \right)$$
$$\leq \frac{1}{2^{n-3}} \Big/ \left(1 - \frac{1}{2} \right), \quad \text{(by Exercise 1.19)}$$
$$= \frac{1}{2^{n-4}}.$$

So if we choose N so that $1/2^{N-4} < \epsilon$, we find that $|a_m - a_n| < \epsilon$ for all $m, n > N$. Thus (a_n), being a Cauchy sequence, converges to a limit α. Finally, using a technique explained before, we let $n \to \infty$ in Eq. (2.12), and deduce that

$$\alpha = \frac{1}{1+\alpha}\,.$$

Thus $\alpha^2 + \alpha - 1 = 0$, and so α, which must be positive, is equal to $(\sqrt{5} - 1)/2$. This number, approximately 0.618, is the so-called "golden number".   □

Remark 2.18

The "golden section" arises when we have three points A, B, C on a line

such that $AB/AC = BC/AB$. If $AC = 1$ and $AB = a$, this means that $a = (1-a)/a$, so that $a^2 + a - 1 = 0$.

EXERCISES

2.22 Show that $((-1)^n/\sqrt{n})$ and $((-1)^n 2^{-n+\frac{1}{2}})$ are Cauchy sequences.

2.23 Let (a_n) be a sequence with the property that, for all $n \geq 2$,

$$|a_{n+1} - a_n| < k|a_n - a_{n-1}|, \qquad (2.13)$$

where $0 < k < 1$. Show that (a_n) is a Cauchy sequence.

Give an example to show that (2.13) is not enough to give a Cauchy sequence if $k = 1$.

2.24 Let (a_n) be a Cauchy sequence, with limit α. Show that if N is such that $|a_m - a_n| < \epsilon$ for all $m > n > N$, then $|\alpha - a_n| \leq \epsilon$ for all $n > N$.

2.25 Let (g_n) be the sequence $(1, 1, 3, 5, 11, 21, 43, \ldots)$, defined by

$$g_1 = g_2 = 1, \quad g_n = g_{n-1} + 2g_{n-2} \quad (n \geq 3),$$

and let $b_n = g_{n+1}/g_n$ $(n = 1, 2, 3, \ldots)$.

a) Show that $b_n b_{n-1} = b_{n-1} + 2$ $(n \geq 2)$.

b) Show that $b_n = 1 + (2/b_{n-1})$ $(n \geq 2)$.

c) Show that $|b_{n+1} - b_n| < (2/3)|b_n - b_{n-1}|$ $(n \geq 2)$.

d) Deduce that (b_n) is a Cauchy sequence, and find its limit.

2.26 Let (L_n) be the sequence given by

$$L_n = \frac{1}{n+1} + \frac{1}{n+1} + \cdots + \frac{1}{2n}.$$

Show that (L_n) is monotonic increasing and that 1 is an upper bound, and deduce that (L_n) has a limit L, where $1/2 \leq L \leq 1$. [In fact, as we shall see in Chapter 6, $L = \log 2 \approx 0.693$.]

2.5 Series

In ordinary usage the words "sequence" and "series" have very much the same meaning. Within mathematics, however, both are technical terms, and their meanings, while related, are different. Given a sequence (a_n), we can define a sequence (S_n), where

$$S_1 = a_1, \quad S_2 = a_1 + a_2, \quad \ldots, \quad S_n = a_1 + a_2 + \cdots + a_n.$$

If the sequence S_n converges, to a limit S (say), then we say that **the (infinite) series $\sum_{n=1}^{\infty} a_n$ converges, and that its sum (to infinity) is** S. Notice that with a series $\sum_{n=1}^{\infty} a_n$ we have *two* associated sequences, namely (a_n), the sequence of **terms** of the series, and (S_n), the sequence of **partial sums** of the series.

An example may help at this point. Consider the sequence $(1/n(n+1))$, and the associated series

$$\sum_{n=1}^{\infty} \frac{1}{n(n+1)} = \frac{1}{1.2} + \frac{1}{2.3} + \frac{1}{3.4} + \cdots .$$

The partial sum

$$S_n = \frac{1}{1.2} + \frac{1}{2.3} + \cdots + \frac{1}{(n-1)n} + \frac{1}{n(n+1)}$$

can be rewritten as

$$\left(1 - \frac{1}{2}\right) + \left(\frac{1}{2} - \frac{1}{3}\right) + \cdots + \left(\frac{1}{n-1} - \frac{1}{n}\right) + \left(\frac{1}{n} - \frac{1}{n+1}\right).$$

Now, in this expression all but the first and last terms cancel, and so

$$S_n = 1 - \frac{1}{n+1} .$$

It is clear that $(S_n) \to 1$, and so we can assert that $\sum_{n=1}^{\infty} 1/n(n+1)$ is a convergent series, whose sum is equal to 1.

We shall want to refer to this example later, and so we record the result:

Theorem 2.19

The series $\sum_{n=1}^{\infty} 1/n(n+1)$ is convergent. Its sum is 1.

Notice that in this example the sequence $(1/n(n+1))$ of *terms* of the series has limit 0. This is necessarily the case for a convergent series:

Theorem 2.20

Let $\sum_{n=1}^{\infty} a_n$ be a convergent series. Then $(a_n) \to 0$.

Proof

Denote the sequence of partial sums by (S_n), and note that

$$a_n = S_n - S_{n-1} \quad (n \geq 2) . \tag{2.14}$$

Since the series is convergent, (S_n) has a limit S, and so, from (2.14), $a_n \to S - S = 0$ as $n \to \infty$. $\square$

It would be pleasant in a way if the converse of this result were also true, but much of the interest and fascination of infinite series arises from the fact that it is not true. The classical example is that of the so-called **harmonic series**

$$\sum_{n=1}^{\infty} \frac{1}{n} = 1 + \frac{1}{2} + \frac{1}{3} + \frac{1}{4} + \cdots.$$

To show that this series is **divergent**, by which we mean of course that its sequence of partial sums is divergent, we shall show that the partial sums can be made arbitrarily large by taking sufficiently many terms. Let us consider the sum to 2^k terms, and write it as

$$S_{2^k} = 1 + \frac{1}{2} + \left(\frac{1}{3} + \frac{1}{4}\right) + \left(\frac{1}{5} + \frac{1}{6} + \frac{1}{7} + \frac{1}{8}\right) + \left(\frac{1}{9} + \frac{1}{10} + \ldots + \frac{1}{16}\right)$$

$$+ \cdots + \left(\frac{1}{2^{k-1} + 1} + \frac{1}{2^{k-1} + 2} + \ldots + \frac{1}{2^k}\right). \qquad (2.15)$$

A typical bracketed expression

$$T_j = \frac{1}{2^{j-1} + 1} + \frac{1}{2^{j-1} + 2} + \ldots + \frac{1}{2^j} \quad (j = 2, \ldots, k)$$

has $2^j - 2^{j-1} = 2^{j-1}(2 - 1) = 2^{j-1}$ terms, each of which is greater than or equal to $1/2^j$. Hence $T_j > 2^{j-1}/2^j = 1/2$ for all j. Hence, from (2.15), we have

$$S_{2^k} = 1 + \frac{1}{2} + T_2 + \ldots + T_k > 1 + \frac{k}{2}.$$

Now, for an arbitrarily chosen positive number M, we can ensure that $1 + (k/2) > M$ by taking $k > 2M - 1$. We deduce that $S_n > M$ whenever $n > 2^{2M-1}$. We have shown that the harmonic series is divergent.

Notice how slowly the series diverges. To ensure that S_n exceeds 100, we require that $n > 2^{199} \approx 8 \times 10^{59}$. Actually it is not quite as bad as that: a more refined argument, given in a later chapter (Remark 6.7), will show that 2.7×10^{43} is enough. But that is still a lot of counting!

We shall see soon that the harmonic series is not a single "freak", but an example of a type of series that is quite common.

We end this section by considering the **geometric series** $\sum_{n=1}^{\infty} ar^{n-1}$. From Exercise 1.10 we have that (for $r \neq 1$)

$$a + ar + ar^2 + \cdots + ar^{n-1} = \frac{a(1 - r^n)}{1 - r} = \frac{a}{1 - r} - \frac{ar^n}{1 - r}.$$

The second term on the right tends to 0 as $n \to \infty$ if and only if $|r| < 1$, and so we have the following result:

Theorem 2.21

The geometric series $\sum_{n=1}^{\infty} ar^{n-1}$ is convergent if and only if $|r| < 1$. The sum is $a/(1-r)$.

EXERCISES

2.27 Determine the sum of the series $\sum_{n=1}^{\infty} 1/n(n+2)$.

2.28 Show that, for all $n \geq 1$,

$$\frac{n}{(n+1)!} = \frac{1}{n!} - \frac{1}{(n+1)!} \, .$$

Hence obtain the sum of the series $\sum_{n=1}^{\infty} [n/(n+1)!]$.

Deduce that $2 \leq \sum_{n=0}^{\infty} (1/n!) \leq 3$.

2.29 In Exercise 2.23 you were asked to show that the condition

$$|a_{n+1} - a_n| < |a_n - a_{n-1}| \quad (n \geq 1)$$

is not sufficient to ensure the convergence of (a_n). Show now that the condition

$$\lim_{n \to \infty} |a_{n+1} - a_n| = 0$$

is not sufficient.

2.6 The Comparison Test

It is useful to begin the study of infinite series by considering series $\sum_{n=1}^{\infty} a_n$ for which all the terms a_n are positive. In such a case we have the advantage that the associated sequence (S_n) is monotonic increasing. The crucial result is the **comparison test**, which compares an unknown series $\sum_{n=1}^{\infty} x_n$ with a known series $\sum_{n=1}^{\infty} a_n$. The simplest form of the test is as follows:

Theorem 2.22 (The Comparison Test)

Let $\sum_{n=1}^{\infty} x_n$ and $\sum_{n=1}^{\infty} a_n$ be series of positive terms.

(i) If $\sum_{n=1}^{\infty} a_n$ is convergent and if $x_n \leq a_n$ for all $n \geq 1$, then $\sum_{n=1}^{\infty} x_n$ is also convergent.

(ii) If $\sum_{n=1}^{\infty} a_n$ is divergent and if $x_n \geq a_n$ for all $n \geq 1$, then $\sum_{n=1}^{\infty} x_n$ is also divergent.

Proof

(i) Denote the partial sums of the series $\sum_{n=1}^{\infty} a_n$ by A_n and of the series $\sum_{n=1}^{\infty} x_n$ by X_n. We are assuming that (A_n) has a limit, which we can denote by A, and since (A_n) is monotonic increasing, we have $A_n \leq A$ for all n. From the assumption that $x_n \leq a_n$ for all n we clearly have that $X_n \leq A_n$ for all $n \geq 1$, and thus $X_n \leq A$. Since (X_n) is monotonic increasing, it follows from Theorem 2.9 that (X_n) is convergent, which is exactly what we require.

(ii) Suppose that this is not so; that is, suppose that $\sum_{n=1}^{\infty} x_n$ is convergent. Since $a_n \leq x_n$ for all n, we deduce from (i) above that $\sum_{n=1}^{\infty} a_n$ is convergent, and we have a contradiction.   □

It is often useful to employ a generalised version of the test. First, it is clear that it is the **ultimate** comparison of the two series that counts and that it is sufficient, in (i) for example, to require that there exists $N \geq 1$ such that $x_n \leq a_n$ for all $n \geq N$. Also, for any positive constant k, the convergence of $\sum_{n=1}^{\infty} a_n$ to a sum A implies the convergence of $\sum_{n=1}^{\infty} ka_n$ to a sum kA (see Exercise 2.30), and so we get a more general version of the comparison test as follows:

Theorem 2.23 (The Comparison Test)

Let $\sum_{n=1}^{\infty} x_n$ and $\sum_{n=1}^{\infty} a_n$ be series of positive terms.

(i) If $\sum_{n=1}^{\infty} a_n$ is convergent and if there exist a positive integer N and a positive real number k such that $x_n \leq ka_n$ for all $n \geq N$, then $\sum_{n=1}^{\infty} x_n$ is also convergent.

(ii) If $\sum_{n=1}^{\infty} a_n$ is divergent and if there exist a positive integer N and a positive real number k such that $x_n \geq ka_n$ for all $n \geq N$, then $\sum_{n=1}^{\infty} x_n$ is also divergent.

The obvious disadvantage of the comparison test is that it demands a "known" series. It is, however, a powerful tool, as the following example shows.

Example 2.24

Show that the series $\sum_{n=1}^{\infty} 1/n^2$ is convergent.

Solution

First, notice that, for all $n \geq 1$, $n^2 \geq n(n+1)/2$. Hence

$$\frac{1}{n^2} \leq \frac{2}{n(n+1)} . \tag{2.16}$$

Now, the series $\sum_{n=1}^{\infty} 1/n(n+1)$ is convergent, as recorded in Theorem 2.19, and so by the comparison test $\sum_{n=1}^{\infty} 1/n^2$ also converges. □

Notice that we do not know the value[5] of $\sum_{n=1}^{\infty} 1/n^2$. The technique does, however, give upper and lower bounds. We have, for all $n \geq 1$,

$$\frac{1}{n(n+1)} \leq \frac{1}{n^2} \leq \frac{2}{n(n+1)} .$$

Since $\sum_{n=1}^{\infty} 1/n(n+1) = 1$, it follows (see Exercise 2.32 below) that $1 \leq \sum_{n=1}^{\infty} 1/n^2 \leq 2$.

More generally, we have the following result:

Theorem 2.25

Let α be a real number.[6] Then $\sum_{n=1}^{\infty} 1/n^{\alpha}$ is convergent if $\alpha > 1$ and is divergent if $\alpha \leq 1$.

Proof

If $\alpha \geq 2$, then $1/n^{\alpha} \leq 1/n^2$ for all $n \geq 1$, and so $\sum_{n=1}^{\infty} 1/n^{\alpha}$ is convergent by the comparison test. If $\alpha \leq 1$, then $1/n^{\alpha} \geq 1/n$ for all $n \geq 1$, and so $\sum_{n=1}^{\infty} 1/n^{\alpha}$ is divergent, again by the comparison test.

It remains to consider the case where $1 < \alpha < 2$. This will follow easily from the **integral test**, to be discussed in Chapter 5. The reader may regard the following elementary, but quite complicated argument as an optional extra. It is a variant of the argument used to establish the divergence of the harmonic series.

For all $N \geq 1$ denote $\sum_{n=1}^{N} 1/n^{\alpha}$ by S_N, and note that, for all $K \geq 1$,

$$S_{2^K} - S_{2^{K-1}} = \frac{1}{(2^{K-1}+1)^{\alpha}} + \frac{1}{(2^{K-1}+2)^{\alpha}} + \cdots + \frac{1}{2^{K\alpha}} .$$

---

[5] In fact, the value of the sum is $\pi^2/6$, which is approximately 1.645, but we cannot prove that by elementary methods.

[6] The careful reader will notice that at this stage we know what n^{α} means if $\alpha = p/q$ is rational – it means $\sqrt[q]{n^p}$, which we can define as $\sup \{x \in \mathbb{R} : x^q < n^p\}$. However, we do not know at all what it means if α is irrational. To give a proper definition we require the exponential function, which we shall encounter in Chapter 6.

There are $2^K - 2^{K-1} = 2^{K-1}$ terms in this expression, each less than $1/2^{(K-1)\alpha}$. Hence

$$S_{2^K} - S_{2^{K-1}} < \frac{2^{K-1}}{2^{(K-1)\alpha}} = \frac{1}{2^{(K-1)(\alpha-1)}}.$$

Now,

$$S_{2^K} = (S_{2^K} - S_{2^{K-1}}) + (S_{2^{K-1}} - S_{2^{K-2}}) + \cdots + (S_2 - S_1) + S_1$$
$$< 1 + 1 + \frac{1}{2^{\alpha-1}} + \frac{1}{2^{2(\alpha-1)}} + \cdots + \frac{1}{2^{(K-1)(\alpha-1)}}$$
$$< 1 + \frac{1}{1 - (1/2^{\alpha-1})} = M \text{ (say)}.$$

For each $N \geq 1$ there exists $K \geq 1$ such that $N \leq 2^K$, and so, since the terms of the series are positive,

$$S_N \leq S_{2^K} < M.$$

We now know that the sequence (S_N) is both monotonic increasing and bounded above. Hence, by Theorem 2.9, (S_N) is convergent. □

The use of the comparison test requires a certain facility in dealing with inequalities. The following examples illustrate the techniques involved.

Example 2.26

Investigate the convergence of $\sum_{n=2}^{\infty} a_n$, where

$$a_n = \frac{n+1}{n^3 - n - 1},$$

Solution

For large n we expect that a_n will not differ much from $1/n^2$, and so we expect that the series will converge. More precisely, since $n + 1 < n^3/2$ for all $n \geq 2$, we have that

$$\frac{n+1}{n^3 - n - 1} < \frac{2n}{n^3 - n - 1} < \frac{2n}{n^3 - (n^3/2)} = \frac{4}{n^2}.$$

Hence $\sum_{n=2}^{\infty} a_n$ is convergent, by comparison with $\sum_{n=2}^{\infty} 1/n^2$. □

Example 2.27

Investigate the convergence of $\sum_{n=3}^{\infty} a_n$, where

$$a_n = \frac{n-2}{n^2 + 2n + 4},$$

Solution

For large n we expect that a_n will not differ much from $1/n$, and so we expect the series to be divergent. More precisely, we notice first that

$$n^2 + 2n + 4 \leq n^2 + 2n^2 + 4n^2 = 7n^2$$

for all $n \geq 1$. Also, $n - 2 \geq n/2$ for all $n \geq 4$. So, for all $n \geq 4$,

$$a_n = \frac{n-2}{n^2 + 2n + 4} \geq \frac{n/2}{n^2 + 2n + 4} \geq \frac{n/2}{7n^2} = \frac{1}{14n},$$

and hence $\sum_{n=3}^{\infty} a_n$ is divergent, by comparison with the harmonic series. $\square$

Actually, one can avoid these technical manipulations with inequalities by developing the theory a little further. If we have two positive sequences (x_n) and (a_n), where we usually want to think of one of them, say (a_n) as "known", we write $x_n \asymp a_n$ (read "x_n has the same order of magnitude as a_n") if $\lim_{n\to\infty}(x_n/a_n) = K$, for some positive constant K. Thus, for example, $\sqrt{n^2 + 1} \asymp 2n$, since

$$\frac{\sqrt{n^2 + 1}}{2n} = \frac{1}{2}\sqrt{1 + (1/n^2)} \to \frac{1}{2} \text{ as } n \to \infty.$$

By contrast, $n^2 + 1 \not\asymp 2n$, since

$$\frac{n^2 + 1}{2n} = \frac{1}{2}(n + (1/n)) \to \infty \text{ as } n \to \infty.$$

A related notation, which we may as well record as this point, gives a stronger relationship between the two sequences: we write $x_n \sim a_n$ (read "x_n is asymptotically equal to a_n") if $\lim_{n\to\infty}(x_n/a_n) = 1$. (This is simply the definition for $x_n \asymp a_n$ in the case where $K = 1$.)

We now have an alternative version of the comparison test:

Theorem 2.28

Let $\sum_{n=1}^{\infty} a_n$, $\sum_{n=1}^{\infty} x_n$ be series of positive terms, and suppose that $x_n \asymp a_n$. Then:

(i) if $\sum_{n=1}^{\infty} a_n$ is convergent, so is $\sum_{n=1}^{\infty} x_n$;

(ii) if $\sum_{n=1}^{\infty} a_n$ is divergent, so is $\sum_{n=1}^{\infty} x_n$.

Proof

We are assuming that

$$\lim_{n\to\infty}(x_n/a_n) = K > 0.$$

Choosing $\epsilon = K/2$, we can therefore assert that there exists N such that

$$\left| \frac{x_n}{a_n} - K \right| < \frac{K}{2}$$

for all $n > N$. Thus

$$\frac{K}{2} < \frac{x_n}{a_n} < \frac{3K}{2}$$

for all $n > N$. Hence $x_n < (3K/2)a_n$ and $x_n > (K/2)a_n$ for all $n > N$. From the first of these inequalities we deduce from Theorem 2.23 that if $\sum_{n=1}^{\infty} a_n$ is convergent, then so is $\sum_{n=1}^{\infty} x_n$, and from the second inequality it similarly follows that if $\sum_{n=1}^{\infty} a_n$ is divergent, then so is $\sum_{n=1}^{\infty} x_n$. □

This version of the comparison test is easier to use.

Example 2.29

Investigate the convergence of $\sum_{n=1}^{\infty} x_n$, where

$$x_n = \frac{n^2 + n + 2}{2n^3 + 3n + 4}.$$

Solution

The dominant terms in the numerator and denominator are n^2 and $2n^3$, respectively, and so we compare the given series with $\sum_{n=1}^{\infty} (n^2/n^3) = \sum_{n=1}^{\infty} (1/n)$:

$$\frac{n^2 + n + 2}{2n^3 + 3n + 4} \bigg/ \frac{1}{n} = \frac{n^3 + n^2 + 2n}{2n^3 + 3n + 4} = \frac{1 + (1/n) + (2/n^2)}{2 + (3/n^2) + (4/n^3)} \to \frac{1}{2}$$

as $n \to \infty$. Hence $x_n \asymp 1/n$, and so $\sum_{n=1}^{\infty} x_n$ is divergent. □

Example 2.30

Investigate the convergence of $\sum_{n=1}^{\infty} x_n$, where

$$x_n = \frac{\sqrt{2n^3 + 1}}{n^3 + 5}.$$

Solution

Here a comparison of the dominant terms suggests a comparison with $\sum_{n=1}^{\infty} (1/n^{3/2})$. Since

$$\frac{\sqrt{2n^3 + 1}}{n^3 + 5} \bigg/ \frac{1}{n^{3/2}} = \frac{\sqrt{2 + (1/n^3)}}{1 + (5/n^3)} \to \sqrt{2}$$

as $n \to \infty$, we have that $x_n \asymp 1/n^{3/2}$, and so $\sum_{n=1}^{\infty} x_n$ is convergent. $\qquad\square$

By using the geometric series as a comparator, we obtain the next result:

Theorem 2.31 (The Ratio Test)

Let $\sum_{n=1}^{\infty} a_n$ be a series of positive terms. Then:

(i) if $\lim_{n\to\infty}(a_{n+1}/a_n) < 1$, the series converges;

(ii) if $\lim_{n\to\infty}(a_{n+1}/a_n) > 1$, or if $(a_{n+1}/a_n) \to \infty$, the series diverges.

Proof

(i) Suppose that $\lim_{n\to\infty}(a_{n+1}/a_n) = l$, where $l < 1$. Choose $\epsilon > 0$ so that $l + \epsilon < 1$. Then there exists N such that

$$l - \epsilon < \frac{a_{n+1}}{a_n} < l + \epsilon < 1$$

for all $n > N$. Denote $l + \epsilon$ by r. Then $a_{n+1} < ra_n$ for all $n \geq N + 1$. That is,

$$a_{N+2} < ra_{N+1}, \; a_{N+3} < ra_{N+2} < r^2 a_{N+1}, \text{etc.},$$

and in general $a_n < a_{N+1}r^{n-N-1}$ for all $n > N$. We thus have a comparison, from the $(N+1)$th term onwards, with the convergent geometric series $\sum_{n=N+1}^{\infty} a_{N+1}r^{n-N-1}$, and so the series $\sum_{n=1}^{\infty} a_n$ is convergent.

(ii) The proof of the second part is similar in spirit, and appears as Exercise 2.39 below. $\qquad\square$

Remark 2.32

It is crucial to note that the ratio test **gives no information at all** if $\lim_{n\to\infty}(a_{n+1}/a_n) = 1$. In both $\sum_{n=1}^{\infty} 1/n$ and $\sum_{n=1}^{\infty} 1/n^2$ we easily see that $\lim_{n\to\infty}(a_{n+1}/a_n) = 1$, but the former series diverges and the latter series converges.

Remark 2.33

If $\lim_{n\to\infty}(a_{n+1}/a_n) > 1$, then $(a_n) \to \infty$.

Remark 2.34

By now it will be clear that the behaviour of an infinite series depends on the "ultimate" nature of its terms. When we apply the comparison test, for example, what the first ten, or one hundred, or one million terms are like is of no consequence: it is the behaviour for **sufficiently large** n that matters. Though we have not emphasised this, the same applies to series of positive terms. All our results apply to **ultimately positive** series, by which of course we mean series $\sum_{n=1}^{\infty} a_n$ for which there exists a positive integer N with the property that $a_n \geq 0$ for all $n > N$.

Example 2.35

Investigate the convergence of

$$\sum_{n=0}^{\infty} \frac{x^n}{n!}.$$

Solution

Here

$$\frac{a_{n+1}}{a_n} = \frac{x^{n+1} n!}{(n+1)! x^n} = \frac{x}{n+1} \to 0 \text{ as } n \to \infty,$$

and so the series converges for all values of x.

It is an incidental consequence of this, following from Theorem 2.20, that $\lim_{n\to\infty}(x^n/n!) = 0$ for all values of x. $\square$

EXERCISES

2.30 Let $\sum_{n=1}^{\infty} a_n$ and $\sum_{n=1}^{\infty} b_n$ be convergent series with sums A and B respectively. Show that $\sum_{n=1}^{\infty}(a_n + b_n) = A + B$ and, for every constant k, $\sum_{n=1}^{\infty}(ka_n) = kA$.

2.31 Are the following statements true or false? If true, give a proof; if false, give a counterexample. The numbers a_n and b_n are positive.

a) If, for all $n \geq 1$, $a_{n+1}/a_n < 1$, then $\sum_{n=1}^{\infty} a_n$ is convergent.

b) If $\lim_{n\to\infty}(a_n - b_n) = 0$ and $\sum_{n=1}^{\infty} b_n$ converges, then $\sum_{n=1}^{\infty} a_n$ converges.

c) $\lim_{n\to\infty}(a_n/b_n) = 1$ and $\sum_{n=1}^{\infty} b_n$ converges, then $\sum_{n=1}^{\infty} a_n$ converges.

d) If $\sum_{n=1}^{\infty} a_n$ converges, then so does $\sum_{n=1}^{\infty} a_n^2$.

e) If $\sum_{n=1}^{\infty} a_n^2$ converges, then so does $\sum_{n=1}^{\infty} a_n$.

f) If $\lim_{n\to\infty}(a_{n+1} + a_{n+2} + \cdots + a_{2n}) = 0$, then $\sum_{n=1}^{\infty} a_n$ is convergent.

2.32 Let $\sum_{n=1}^{\infty} a_n$ and $\sum_{n=1}^{\infty} b_n$ be convergent series with sums A and B respectively. Show that, if $a_n \leq b_n$ for all $n \geq 1$, then $A \leq B$.

2.33 Investigate the convergence of

$$\sum_{n=1}^{\infty} \frac{\sqrt{n+1}}{n^2+2}, \quad \sum_{n=2}^{\infty} \frac{n+3}{n^4-1}, \quad \sum_{n=1}^{\infty} \frac{n+1}{\sqrt{n^3+2}}.$$

2.34 Investigate the convergence of

$$\sum_{n=1}^{\infty} (\sqrt{n+1} - \sqrt{n}), \quad \sum_{n=1}^{\infty} \left(\frac{1}{\sqrt{n}} - \frac{1}{\sqrt{n+1}} \right).$$

2.35 Investigate the convergence of

$$\sum_{n=1}^{\infty} \frac{n!}{n^n}, \quad \sum_{n=1}^{\infty} \frac{n^3}{n!}.$$

2.36 Show that, if $\sum_{n=1}^{\infty} a_n$ and $\sum_{n=1}^{\infty} b_n$ are both convergent series of non-negative terms, then so is $\sum_{n=1}^{\infty} \max \{a_n, b_n\}$.

2.37 Let $\sum_{n=1}^{\infty} a_n$ and $\sum_{n=1}^{\infty} b_n$ be convergent series of non-negative terms. Show that $\sum_{n=1}^{\infty} (a_n b_n)^{1/2}$ is convergent. Give an example to show that the converse implication is false.

2.38 Give an example of two divergent series $\sum_{n=1}^{\infty} a_n$ and $\sum_{n=1}^{\infty} b_n$ of positive terms with the property that $\sum_{n=1}^{\infty} \min \{a_n, b_n\}$ is convergent.

2.39 Prove the second part of Theorem 2.31.

2.40 Use the ratio test to show that, for each fixed k and each a such that $0 < a < 1$ the series $\sum_{n=1}^{\infty} n^k a^n$ is convergent. Deduce that $\lim_{n \to \infty} n^k a^n = 0$.

2.7 Series of Positive and Negative Terms

We turn now to "mixed" series $\sum_{n=1}^{\infty} a_n$, in which there may be a mixture of positive and negative terms. Such a series is called **absolutely convergent** if the corresponding series $\sum_{n=1}^{\infty} |a_n|$ is convergent. For example, the series

$$\sum_{n=1}^{\infty} \frac{(-1)^{n-1}}{n^2} = 1 - \frac{1}{2^2} + \frac{1}{3^2} - \frac{1}{4^2} + \cdots \tag{2.17}$$

is absolutely convergent, since $\sum_{n=1}^{\infty} 1/n^2$ is convergent. The following theorem enables us to conclude that the series (2.17) is convergent:

Theorem 2.36

Every absolutely convergent series is convergent.

Proof

Let $\sum_{n=1}^{\infty} |a_n|$ be convergent, and, as usual, let

$$S_n = \sum_{r=1}^{n} |a_r|$$

denote the sum of the first n terms. Since (S_n) converges, it must be a Cauchy sequence, and so, for all $\epsilon > 0$ there exists N such that $|S_m - S_n| < \epsilon$ for all $m > n > N$. Let T_n be the sum of the first n terms of the series $\sum_{n=1}^{\infty} a_n$. Then, for all $m > n > N$,

$$\begin{aligned}
|T_m - T_n| &= |a_{n+1} + a_{n+2} + \cdots + a_m| \\
&\leq |a_{n+1}| + |a_{n+2}| + \cdots + |a_m| \\
&= |S_m - S_n| < \epsilon.
\end{aligned}$$

Thus (T_n) is a Cauchy sequence, and so is convergent by Theorem 2.16.   □

If a series $\sum_{n=1}^{\infty} a_n$ is convergent but not absolutely convergent – and we shall see shortly that this is possible – we say that it is **conditionally convergent**. Many of the series we encounter in practice are what we call **alternating series,** by which we mean that the terms are alternately positive and negative. For such series there is a useful test, called the **Leibniz test.**[7]

Theorem 2.37 (The Leibniz Test)

Let (a_n) be a sequence of positive terms, and suppose that

(i) (a_n) is monotonic decreasing;

(ii) $(a_n) \to 0$.

[7] Gottfried Wilhelm Leibniz (1646–1716) was, with Isaac Newton (1643–1727), one of the founders of the calculus.

Then the alternating series

$$\sum_{n=1}^{\infty}(-1)^{n-1}a_n = a_1 - a_2 + a_3 - \cdots$$

is convergent.

Proof

Consider, for each $N \geq 1$, the sum of the series to $2N$ terms:

$$S_{2N} = (a_1 - a_2) + (a_3 - a_4) + \cdots + (a_{2N-1} - a_{2N}).$$

Each of the brackets is non-negative, since (a_n) is monotonic decreasing, and so the sequence $(S_{2N}) = (S_2, S_4, S_6, \ldots)$ is monotonic increasing. Similarly

$$S_{2N+1} = a_1 - (a_2 - a_3) - (a_4 - a_5) - \cdots - (a_{2N} - a_{2N+1}).$$

Again each of the brackets is positive, and so the sequence

$$(S_{2N+1}) = (S_1, S_3, S_5, \ldots)$$

is monotonic decreasing.

Notice next that

$$S_{2N+1} = S_{2N} + a_{2N+1} \geq S_{2N},$$

so that, for all $N \geq 1$

$$S_2 \leq S_4 \leq \cdots \leq S_{2N} \leq S_{2N+1} \leq \cdots \leq S_3 \leq S_1.$$

It follows that (S_{2N}), being a monotonic increasing sequence bounded above by S_1, has a limit S_E (say), while (S_{2N+1}), being a monotonic decreasing sequence bounded below by S_2, has a limit S_O.

Now let $N \to \infty$ in the equality $S_{2N+1} = S_{2N} + a_{2N+1}$. We obtain $S_O = S_E + 0$, and so $S_O = S_E, = S$ (say).

At this point you may regard it as obvious that $(S_n) \to S$. If so, read no further. A formal proof is as follows. Let $\epsilon > 0$. Then, since $(S_{2N}) \to S$, there exists M_1 such that $|S_{2N} - S| < \epsilon$ for all $N > M_1$. Similarly, there exists M_2 such that $|S_{2N+1} - S| < \epsilon$ for all $N > M_2$. Let $M_3 = \max\{2M_1, 2M_2 + 1\}$, and let $n > M_3$. If n is even, say $n = 2k$, then $k > M_1$ and so $|S_n - S| < \epsilon$, while if n is odd, say $n = 2k + 1$, then $k > M_2$ and so $|S_n - S| < \epsilon$. Thus $|S_n - S| < \epsilon$ for all $n > M_3$, and so $(S_n) \to S$. $\qquad\square$

The Leibniz test applies in particular to the **alternating harmonic series**

$$\sum_{n=1}^{\infty} \frac{(-1)^{n-1}}{n} = 1 - \frac{1}{2} + \frac{1}{3} - \cdots,$$

This series is convergent, but is *not* absolutely convergent by virtue of our observations regarding the harmonic series (page 49).

EXERCISES

2.41 Investigate the convergence of the series

$$\sum_{n=1}^{\infty} \frac{(-1)^{n-1}}{2n-1}, \quad \sum_{n=1}^{\infty} \frac{(-1)^{n-1}}{n^{3/2}}.$$

2.42 Give an example of a divergent alternating series satisfying only Condition (ii) of the Leibniz test.

2.43 A celebrated theorem due to Riemann[8] shows that a conditionally convergent series can be rearranged so as to sum to any real number, or to diverge to ∞, or to diverge to $-\infty$. This exercise has the more modest aim of showing that a rearrangement may have a different sum. Consider the alternating harmonic series

$$1 - \frac{1}{2} + \frac{1}{3} - \cdots,$$

with sum S, and denote its sum to n terms ($n = 1, 2, 3, \ldots$) by S_n. Consider also the rearranged series

$$1 + \frac{1}{3} - \frac{1}{2} + \frac{1}{5} + \frac{1}{7} - \frac{1}{4} + \frac{1}{9} + \frac{1}{11} - \frac{1}{6} + \cdots,$$

and denote its sum to n terms by T_n. For each $n \geq 1$, let

$$H_n = 1 + \frac{1}{2} + \frac{1}{3} + \cdots + \frac{1}{n}.$$

a) Show that $S_{2n} = H_{2n} - H_n$ for all $n \geq 1$.

b) Show that

$$T_{3n} = H_{4n} - \frac{1}{2}H_{2n} - H_n = S_{4n} + \frac{1}{2}S_{2n},$$

and deduce that the rearranged series has sum $3S/2$.

---

[8] Georg Friedrich Bernhard Riemann, 1826–1866

2.44 With the same notation as in the previous exercise, show that the series

$$1 - \frac{1}{2} - \frac{1}{4} + \frac{1}{3} - \frac{1}{6} - \frac{1}{8} + \frac{1}{5} - \frac{1}{10} - \frac{1}{12} + \cdots$$

has sum $S/2$.

3
Functions and Continuity

3.1 Functions, Graphs

The notion of function is fundamental to both pure and applied mathematics. To give the "posh" definition first, if A and B are non-empty sets, a **function** f from A into B (usually written $f : A \to B$) is defined as a subset f of the Cartesian product $A \times B$ with the property that, for all x in A and all y_1, y_2 in B,

$$(x, y_1) \in f \text{ and } (x, y_2) \in f \implies y_1 = y_2 \, .$$

To put it another way, for every x in A, the **domain** of the function, there is a *unique* y in B such that $(x, y) \in f$. In practice we denote this unique y by $f(x)$, and say that $f(x)$ is the **image of x under** f, or the **value of f at** x. We shall sometimes want to refer to the domain A of f as dom f. The set B is sometimes called the **codomain** of f

This change of notation brings in its wake a change of viewpoint, leading as it does to the useful notion of a function as a "process", converting each x to its image $f(x)$. The important thing is to realise that this process cannot necessarily be described by a single formula. In many important cases it can: for example, the formulae

$$f(x) = x^2, \quad f(x) = (x^3 + 1)/(x^4 + 1) \quad (x \in \mathbb{R})$$

both describe functions with domain $\mathbb{R}$. But so also does the formula

$$f(x) = \begin{cases} 0 & \text{if } x \in \mathbb{Q} \\ 1 & \text{if } x \in \mathbb{R} \setminus \mathbb{Q}, \end{cases} \tag{3.1}$$

or the "split" formula

$$f(x) = \begin{cases} x - 1 & \text{if } x \le 2 \\ -3 + 4x - x^2 & \text{if } x > 2. \end{cases} \tag{3.2}$$

The **image** im f of a function $f : A \to B$ is a subset of the codomain B, defined by

$$\operatorname{im} f = \{f(x) : x \in A\}.$$

It may be a proper subset of B.

The definition of a function is very general, and in fact encompasses the notion of a sequence encountered in Chapter 2. A sequence is properly defined as a function $f : \mathbb{N} \to \mathbb{R}$. Such a function is specified by listing its values $f(1), f(2), f(3), \ldots$, and it is in this way that we usually regard it, writing it as the sequence $(f(n))$.

In this chapter we shall be mostly concerned with functions f where dom f is either $\mathbb{R}$ or an interval within $\mathbb{R}$. In such cases it can be useful to draw the **graph** $\{(x, f(x)) : x \in \operatorname{dom} f\}$ of the function: for example, the graph of the function given by formula (3.2) is given in Fig. 3.1.

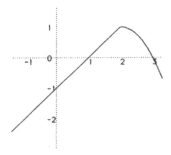

Figure 3.1.

On the other hand, it is not possible to draw a convincing graph of the function given by (3.1).

It is quite important at times to distinguish between a function f and its **value** $f(x)$ at a point x. We shall therefore try to avoid descriptions such as "the function $x^2 + 2$", and instead refer to **the function f given by**

$$f(x) = x^2 + 2 \quad (x \in \mathbb{R}),$$

or, more simply, to **the function** $x \mapsto x^2 + 2$ $(x \in \mathbb{R})$. (Read "x maps to $x^2 + 2$".) We shall omit the reference to the domain if the context allows.

Sometimes a formula creates an automatic restriction on the domain. For example, in the function $x \mapsto \sqrt{1 - x^2}$ the largest possible domain is $[-1, 1]$, since $\sqrt{1-x^2}$ is undefined for $|x| > 1$. The image is $[0,1]$, and the graph is the semicircle given in Fig. 3.2.

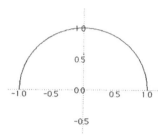

Figure 3.2. The function $x \mapsto \sqrt{1-x^2}$

Similarly, the domain of the function $x \mapsto \sqrt{x}$ cannot be larger than $[0,\infty)$. Its graph is given in Fig. 3.3.

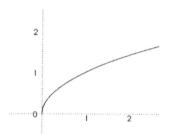

Figure 3.3. The function $x \mapsto \sqrt{x}$

At this stage it is useful to mention some other functions that play a part in future chapters. We have already encountered the function $x \mapsto |x|$. Its image is $[0,\infty)$ and its graph is

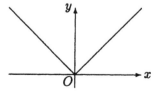

Next, we define
$$\lfloor x \rfloor = \max\{n \in \mathbb{Z} : n \le x\}. \tag{3.3}$$

This is sometimes called the **integral part of** x. The image of the function $x \mapsto \lfloor x \rfloor$ is $\mathbb{Z}$, and the graph is

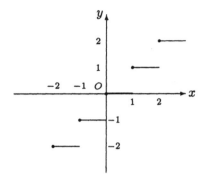

(The dots serve to indicate that the value of the function at each integer n is n.)

EXERCISES

3.1 Let $\lceil x \rceil = \min \{n \in \mathbb{Z} : n \geq x\}$. Sketch the graph of the function $x \mapsto \lceil x \rceil$.

3.2 Sketch the graph of the functions f_1, f_2, f_3, where

a) $f_1(x) = \lceil x \rceil - \lfloor x \rfloor$;

b) $f_2(x) = x - \lfloor x \rfloor$;

c) $f_3(x) = x/|x|$.

3.2 Sums, Products, Compositions; Polynomial and Rational Functions

Let f, g be functions whose domains and images are subsets of $\mathbb{R}$. We write $f = g$ if $\operatorname{dom} f = \operatorname{dom} g$ and $f(x) = g(x)$ for all x in the common domain. We can define a function $f + g$, the **sum** of f and g, by the simple rule that $\operatorname{dom}(f + g) = \operatorname{dom} f \cap \operatorname{dom} g$, and

$$(f + g)(x) = f(x) + g(x) \quad (x \in \operatorname{dom}(f + g)).$$

The **product** $f \cdot g$ of f and g again has domain $\operatorname{dom} f \cap \operatorname{dom} g$, and is given by

$$(f \cdot g)(x) = f(x)g(x) \quad (x \in \operatorname{dom}(f \cdot g)).$$

An important special case arises when we multiply the function f by a **constant** function $x \mapsto k$, where k is an arbitrary real number. We obtain the function kf, given by

$$(kf)(x) = kf(x) \quad (x \in \operatorname{dom} f).$$

In particular, the function $(-1)f$ is usually denoted by $-f$; thus

$$(-f)(x) = -f(x) \quad (x \in \operatorname{dom} f).$$

The function $f - g$ is then defined as $f + (-g)$.

Let us define certain basic functions C_k ($k \in \mathbb{R}$) (the **constant** functions), and i (the **identity** function), both with domain $\mathbb{R}$, as follows:

$$C_k(x) = k, \quad i(x) = x \quad (x \in \mathbb{R}). \tag{3.4}$$

A **polynomial function** p is a function that can be obtained from C_k and i by finitely many applications of addition and multiplication. So, for example, the function p defined by

$$p(x) = x^3 - \sqrt{3}x^2 + \frac{1}{2}x - 2$$

can be written as

$$p = (i \cdot i \cdot i) + (C_{-\sqrt{3}} \cdot i \cdot i) + (C_{1/2} \cdot i) + C_{-2}.$$

The domain of a polynomial function is the whole of $\mathbb{R}$.

We have defined the sum and the product of two functions. We come now to the quotient. First, the **reciprocal** $1/g$ of a function g has domain $\{x \in \operatorname{dom} g : g(x) \neq 0\}$, and is given by

$$\left(\frac{1}{g}\right)(x) = \frac{1}{g(x)} \quad (x \in \operatorname{dom}(1/g)).$$

Then we define f/g to be $f \cdot (1/g)$, the domain being $\operatorname{dom} f \cap \operatorname{dom}(1/g)$.

A **rational function** r is a function obtained from the functions C_k and i (defined above in (3.4)) by finitely many applications of addition, multiplication and division. Such a function can always be expressed as p/q, where p and q are polynomial functions,[1] and its domain is $\{x : q(x) \neq 0\}$.

Let $f : A \to B$ and $g : B \to C$ be functions. The **composition** $g \circ f : A \to C$ is defined by

$$(g \circ f)(a) = g(f(a)) \quad (a \in A).$$

---

[1] That is perhaps not quite obvious. It is explored in Exercise 3.4 at the end of the section.

If f and g are functions whose domains are subsets of $\mathbb{R}$, then we normally take the sensible line that

$$\mathrm{dom}(g \circ f) = \{x \in \mathrm{dom}\, f \,:\, f(x) \in \mathrm{dom}\, g\}.$$

Thus, for example, if $f(x) = 3x + 2$ $(x \in \mathbb{R})$ and $g(x) = \sqrt{1 - x^2}$ $(x \in [-1, 1])$, then

$$\mathrm{dom}(g \circ f) = \{x \,:\, -1 \le 3x + 2 \le 1\} = \{x \,:\, -1 \le x \le -1/3\} = [-1, -1/3],$$

and

$$(g \circ f)(x) = \sqrt{1 - (3x + 2)^2} = \sqrt{-3 - 12x - 9x^2}.$$

From the definition of a polynomial function it is clear that the sum and product of two polynomial functions are polynomial functions, and from this it follows that if f and g are polynomial functions then so is $g \circ f$. For example, if $f(x) = x^2 + 1$ and $g(x) = 2x^3 - x$, then

$$(g \circ f)(x) = g(f(x)) = 2(x^2 + 1)^3 - (x^2 + 1)$$
$$= 2(x^6 + 3x^4 + 3x^2 + 1) - x^2 - 1 = 2x^6 + 6x^4 + 5x^2 + 1.$$

Similarly, if f and g are rational functions, then so is $g \circ f$. (See Exercise 3.6.)

EXERCISES

3.3 In each case find the domain of $g \circ f$ and a formula for $(g \circ f)(x)$.

a) $f(x) = 2x - 3$ $(x \in \mathbb{R})$, $g(x) = \sqrt{x}$ $(x \in [0, \infty))$.

b) $f(x) = \sqrt{4 - x^2}$ $(x \in [-2, 2])$, $g(x) = 1/(x^3 - 1)$ $(x \in \mathbb{R} \setminus \{1\})$.

c) $\mathrm{dom}\, f = \mathrm{dom}\, g = [0, \infty)$, and

$$f(x) = \begin{cases} 1/q & \text{if } x \text{ is rational and is expressed} \\ & \quad \text{as } p/q \text{ in lowest terms} \\ 0 & \text{if } x \text{ is irrational,} \end{cases}$$

$$g(x) = \begin{cases} 1 & \text{if } x \text{ is rational} \\ 0 & \text{if } x \text{ is irrational.} \end{cases}$$

3.4 In building up a rational function from C_k and i, two successive uses of division can always be replaced by a multiplication followed by a division:

$$(f/g)/h = f/(g \cdot h), f/(g/h) = (f \cdot h)/g.$$

(There is a small problem here over the precise domains involved. Strictly speaking, the function $x \mapsto 1/(1/x)$ differs from $x \mapsto x$, in that 0 is absent from the domain of the first function. But changes such as that from $1/(1/x)$ to x can only "improve" the function by adding points to its domain.)

a) Show that a division followed by an addition or subtraction can be replaced by an addition or subtraction followed by a division.

b) Show that a division followed by a multiplication can be replaced by a multiplication followed by a division.

c) Deduce that every rational function is expressible as p/q, where p and q are polynomials.

3.5 Let f, g and h be functions with domain and codomain $\mathbb{R}$. Show that $(f+g) \circ h = (f \circ h) + (g \circ h)$. Show by an example that $f \circ (g+h)$ and $(f \circ g) + (f \circ h)$ may be different.

3.6 Let q, r be rational functions. Show that

$$q + r, \ q \cdot r \text{ and } q/r$$

are all rational functions, and deduce that the composition of two rational functions is a rational function.

3.7 Let i, f, g, h be defined (for $x \neq 0$) by

$$i(x) = x, \quad f(x) = -x, \quad g(x) = \frac{1}{x}, \quad h(x) = -\frac{1}{x}.$$

Show that the compositions of the functions can be summarised in the table

$\circ$	i	f	g	h
i	i	f	g	h
f	f	i	h	g
g	g	h	i	f
h	h	g	f	i

3.8 Draw up a similar table for the six functions given, for x in $\mathbb{R}\setminus\{0,1\}$) by

$$i(x) = x, \quad f(x) = \frac{1}{1-x}, \quad g(x) = 1 - \frac{1}{x},$$

$$p(x) = \frac{1}{x}, \quad q(x) = 1 - x, \quad r(x) = \frac{x}{x-1}.$$

Here, and in the previous example, the functions form a **group** under composition. (See [2].)

3.3 Circular Functions

The assumption behind the plan of this book is that the reader has enjoyed (or at any rate experienced) a standard course in calculus, and in particular has encountered the **circular functions** sin and cos. The name "circular" arises from the following standard definition. Let C be the **unit circle** (that is, the circle with centre $(0, 0)$ and radius 1), and let $P(a, b)$ be a point on the circle.

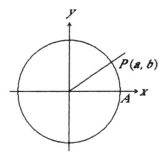

Figure 3.4. The unit circle

See Fig. 3.4. Angles such as $\angle AOP$ are frequently measured in *degrees*, there being 90° in a right angle, but the natural way to measure the angle $\angle AOP$ is by means of the length of the circular arc AP. One **radian** is the angle for which arc $AP = 1$. (It is approximately 57.3°.) Thus a right angle has **circular measure** $\pi/2$ radians.

If P is the point (a, b) and $\angle AOP = \theta$ (measured in radians), then we define $\cos \theta = a$ and $\sin \theta = b$. Notice that from the equation $x^2 + y^2 = 1$ of the circle C we immediately have the standard identity

$$\cos^2 \theta + \sin^2 \theta = 1 \,.$$

We have defined sin and cos for $0 \le \theta < 2\pi$. It is reasonable, however, to extend the meaning. First, we may regard the journey from A to P along the arc as positive if counterclockwise and negative if clockwise. If $P'(a, -b)$ is the reflection of P in the x-axis, then $\angle AOP' = -\theta$, and so

$$\cos(-\theta) = \cos\theta, \quad \sin(-\theta) = -\sin\theta \,. \tag{3.5}$$

This extends the domain of cos and sin to the interval $(-2\pi, 2\pi)$. If $\theta > 2\pi$ we can write $\theta = 2n\pi + \phi$, where n is a positive integer and $0 \le \phi < 2\pi$. The arc length θ corresponds to n complete counterclockwise journeys round the circle, followed by an arc ending at the same point P as the arc length ϕ. So it is reasonable to define

$$\cos(\phi + 2n\pi) = \cos\phi, \quad \sin(\phi + 2n\pi) = \sin\phi \quad (0 \le \phi < 2\pi, \, n \in \mathbb{N}).$$

This, together with (3.5), extends the domains of cos and sin to the whole of $\mathbb{R}$.

We frequently wish to abandon the geometric viewpoint, and simply regard cos and sin as functions with domain $\mathbb{R}$. The graphs are given in Figs. 3.5 and 3.6.

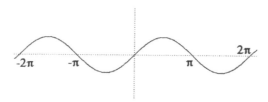

Figure 3.5. The function sin

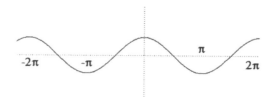

Figure 3.6. The function cos

There is a concealed assumption in all of this, namely that we know what we mean by the length of a curved line, and this is not a trivial matter. In fact we shall not give a "proper" definition of sin and cos until Chapter 8, and logically we should not mention these functions until then. From the point of view of teaching and learning, however, it is rather artificial to "pretend" that one does not know these functions. Our approach will therefore be to take them as provisionally known, pending a strict definition.

For the moment we shall also assume the sum and difference formulae

$$\cos(x \pm y) = \cos x \cos y \mp \sin x \sin y,$$
$$\sin(x \pm y) = \sin x \cos y \pm \cos x \sin y, \tag{3.6}$$

and we shall in particular want to quote two of the straightforward consequences of these formulae:

$$\sin x - \sin y = 2 \cos \frac{x+y}{2} \sin \frac{x-y}{2},$$

$$\cos x - \cos y = -2 \sin \frac{x+y}{2} \sin \frac{x-y}{2}. \qquad (3.7)$$

It is scarcely necessary to remark that once we know sin and cos, we also know the other circular functions:

$$\tan x = \frac{\sin x}{\cos x}, \quad \cot x = \frac{\cos x}{\sin x}, \quad \sec x = \frac{1}{\cos x}, \quad \operatorname{cosec} x = \frac{1}{\sin x},$$

these being defined whenever the denominator is non-zero.

EXERCISES

3.9 Show that $|\cos \theta| \le 1$, $\quad |\sin \theta| \le 1$ for all θ in $\mathbb{R}$.

3.10 Establish the identities:

$$\cos^2 \theta = \frac{1}{2}(1 + \cos 2\theta), \quad \sin^2 \theta = \frac{1}{2}(1 - \cos 2\theta).$$

3.11 Establish the identities:

$$\sin x + \sin y = 2 \sin \frac{x+y}{2} \cos \frac{x-y}{2},$$

$$\cos x + \cos y = 2 \cos \frac{x+y}{2} \cos \frac{x-y}{2}.$$

3.12 Show that

$$\tan(x+y) = \frac{\tan x + \tan y}{1 - \tan x \tan y}.$$

3.13 Let $t = \tan(\theta/2)$. Show that

$$\cos \theta = \frac{1 - t^2}{1 + t^2}, \quad \sin \theta = \frac{2t}{1 + t^2}, \quad \tan \theta = \frac{2t}{1 - t^2}.$$

3.4 Limits

Let $f : A \to \mathbb{R}$ be a function, where $A \subseteq \mathbb{R}$. With an easy modification of our definition for sequences, we can talk of $\lim_{x \to \infty} f(x)$, provided only that the domain A of f contains an interval $[a, \infty)$ for some a in $\mathbb{R}$. We write that $\lim_{x \to \infty} f(x) = l$, **or that $f(x) \to l$ as $x \to \infty$, if, for all $\epsilon > 0$ there exists M in $\mathbb{R}$ such that $|f(x) - l| < \epsilon$ for all $x > M$.** Thus, for example, if we consider the function f given by

$$f(x) = \frac{x+1}{|2x+3|} \quad (x \in \mathbb{R}, \ x \ne -\frac{3}{2}),$$

then

$$\lim_{x \to \infty} f(x) = \frac{1}{2},$$

since, for all $x > -3/2$,

$$\left| \frac{x+1}{|2x+3|} - \frac{1}{2} \right| = \left| \frac{2x+2-2x-3}{2(2x+3)} \right| = \left| \frac{1}{2(2x+3)} \right| < \frac{1}{4x}.$$

Thus, if we define M to be $1/4\epsilon$, we can assert that $|f(x) - \frac{1}{2}| < \epsilon$ whenever $x > M$.

We can also define $\lim_{x \to -\infty} f(x)$ as $\lim_{x \to \infty} f(-x)$. Thus, for example, if f is as above, then

$$f(-x) = \frac{-x+1}{|-2x+3|} = \frac{-x+1}{|2x-3|} = \frac{-x+1}{2x-3}$$

whenever $x > 3/2$. Hence, if $x > 3$ (which implies that $2x - 3 > x$)

$$\left| f(-x) - \left(-\frac{1}{2} \right) \right| = \left| \frac{-x+1}{2x-3} + \frac{1}{2} \right| = \left| \frac{1}{2(2x-3)} \right| < \frac{1}{2x}.$$

Let $\epsilon > 0$ be given. Then, for all $x > M = \max\{3, 1/2\epsilon\}$,

$$\left| f(-x) - \left(-\frac{1}{2} \right) \right| < \epsilon,$$

and so $\lim_{x \to -\infty} f(x) = -1/2$.

A function f defined on an interval I is said to be **strictly increasing if**, for all x, y in I,

$$x < y \implies f(x) < f(y),$$

and **strictly decreasing** if for all x, y in I,

$$x < y \implies f(x) > f(y).$$

If it is either strictly increasing or strictly decreasing it is called **strictly mono-tonic**. If < and > are replaced by ≤ and ≥ in these definitions, then we obtain the definitions of **increasing**, **decreasing** and **monotonic** functions.

A function f is said to be **bounded above** if the set $\{f(x) : x \in I\}$ is bounded above, **bounded below** if the set $\{f(x) : x \in I\}$ is bounded below, and **bounded** if it is bounded both above and below. By analogy with Theorems 2.9 and 2.10, we have

Theorem 3.1

If f, defined on an interval $[a, \infty)$, is an increasing function and is bounded above, then $\lim_{x \to \infty} f(x)$ exists. The limit again exists if f is decreasing and is bounded below.

Proof

The proof is left as an exercise (Exercise 3.14).   □

Again, by analogy with Theorems 2.15 and 2.16, we have what is sometimes referred to as the **General Principle of Convergence**:

Theorem 3.2

Let f be a function with domain $[a, \infty)$. Then $\lim_{x \to \infty} f(x)$ exists if and only if, for every $\epsilon > 0$ there exists X in $[a, \infty)$ such that $|f(x) - f(y)| < \epsilon$ for all $x, y > X$.

Proof

Suppose first that $\lim_{x \to \infty} f(x)$ exists and equals L. Let $\epsilon > 0$ be given. Then there exists X such that $|f(x) - L| < \epsilon/2$ for all $x > X$. Hence, for all $x, y > X$,

$$|f(x) - f(y)| = |[f(x) - L] + [L - f(y)]| \leq |f(x) - L| + |f(y) - L| < \epsilon.$$

Conversely, suppose that, for all $\epsilon > 0$ there exists X such that $|f(x) - f(y)| < \epsilon$ for all $x, y > X$. Then the sequence $(f(n))_{n \in \mathbb{N}}$ is a Cauchy sequence, and so, by Theorem 2.16, has a limit L.

Let $\epsilon > 0$ be given. There exists K_1 such that $|f(n) - L| < \epsilon/2$ for all integers $n > K_1$. Also, by the property we are assuming, there exists K_2 such that $|f(x) - f(y)| < \epsilon/2$ for all $x, y > K_2$. Let $K = \max\{K_1, K_2\}$ and let n be an integer greater than K. Then, for all $x > K$,

$$|f(x) - L| = |[f(x) - f(n)] + [f(n) - L]| \text{ (where } n \in \mathbb{N} \text{ and } n > K)$$

$$\leq |f(x) - f(n)| + |f(n) - L| < \epsilon,$$

and so $\lim_{x \to \infty} f(x) = L$. $\qquad\qquad\qquad\qquad\qquad\qquad\qquad\qquad\qquad\quad\square$

So far, this is very similar to what we have already seen for sequences, but with real functions f we have another possibility. From a calculus course we are familiar with arguments of the following type:

$$\lim_{x \to 1} \frac{x^2 - 1}{x^2 + x - 2} = \lim_{x \to 1} \frac{(x+1)(x-1)}{(x+2)(x-1)} = \lim_{x \to 1} \frac{x+1}{x+2} = \frac{2}{3}. \qquad (3.8)$$

There is a logical difficulty about this kind of argument, for the two functions

$$x \mapsto \frac{x^2 - 1}{x^2 + x - 2} \quad \text{and} \quad x \mapsto \frac{x+1}{x+2}$$

are not in fact quite equal, the first being undefined for $x = 1$. And $x = 1$ is precisely the point that is of interest! We need a definition of this kind of limit in the same style as our previous definitions.

Let f be a real function and let $a \in \mathbb{R}$ (where a may or may not be an element of dom f.) To define the statement

$$\lim_{x \to a} f(x) = l$$

we need to say that $f(x)$ can be made arbitrarily close to l by taking x sufficiently close to a. More precisely, we say that $\lim_{x \to a} f(x) = l$ (in words, the limit as x tends to a of $f(x)$ equals l) if **for all $\epsilon > 0$ there exists $\delta > 0$ such that $|f(x) - l| < \epsilon$ for all x in dom $f \setminus \{a\}$ such that $|x - a| < \delta$**. As before, we can usefully think of ϵ as "small", and we can expect that δ will depend on ϵ.

Let us look again at the example (3.8) in a more precise way. We have a function f with domain $[0, \infty) \setminus \{1\}$, given by

$$f(x) = \frac{x^2 - 1}{x^2 + x - 2}.$$

Let $\epsilon > 0$ be given. Then, noting that $x + 2 \geq 2$ for all $x \geq 0$, we have that, for all x in $[0, \infty) \setminus \{1\}$,

$$\left| f(x) - \frac{2}{3} \right| = \left| \frac{x^2 - 1}{x^2 + x - 2} - \frac{2}{3} \right| = \left| \frac{x+1}{x+2} - \frac{2}{3} \right|$$

$$= \left| \frac{3x + 3 - 2x - 4}{3(x+2)} \right| = \left| \frac{x-1}{3(x+2)} \right| \leq \frac{|x-1|}{6}.$$

It follows that if we define $\delta = 6\epsilon$ we have that $|f(x) - (2/3)| < \epsilon$ for all x in $[0, \infty) \setminus \{1\}$ such that $|x - a| < \delta$.

In the boldface statement above the exclusion of a involved in the use of the phrase "x in dom $f \setminus \{a\}$" is necessary. Suppose that we have a function f defined (admittedly somewhat perversely) by

$$f(x) = \begin{cases} x^2 & (x \in \mathbb{R}, \ x \neq 3) \\ 0 & (x = 3). \end{cases} \qquad (3.9)$$

Then $\lim_{x \to 3} f(x) = 9$. To see this, suppose that $|x - 3| < 1$. Then $2 < x < 4$ and so $5 < x + 3 < 7$. Suppose also that $x \neq 3$. Then

$$|f(x) - 9| = |x + 3||x + 3| < 7|x - 3|. \qquad (3.10)$$

Let $\epsilon > 0$ be given. Then $|f(x) - 9| < \epsilon$ for all x in $\mathbb{R} \setminus \{3\}$ such that $|x - 3| < \min\{1, \epsilon/7\}$. Had we not excluded 3 from this last statement, it would have been untrue: $|f(3) - 9| = 9$.

The technicalities of this kind of argument can be a little awkward, and it must be emphasised that the seemingly arbitrary statement above, "Suppose that $|x - 3| < 1$" is exactly as it seems – arbitrary! Something of the kind is needed to get an inequality of the general type of (3.10), but we could just as well have said "Suppose that $|x - 3| < 15$", obtaining $|f(x) - 9| < 18|x - 3|$, and then taking $\delta = \min\{15, \epsilon/18\}$. Fortunately, as we now see, there are some general theorems about limits that reduce the need for arguments from first principles. The proofs are very similar in spirit to those in Theorem 2.8.

Given a function f, we define the function $|f|$ by

$$|f|(x) = |f(x)| \quad (x \in \text{dom } f). \qquad (3.11)$$

Theorem 3.3

Let f, g be real functions, and suppose that $\lim_{x \to a} f(x) = l$, $\lim_{x \to a} g(x) = m$. Then:

(i) $\lim_{x \to a} |f|(x) = |l|$;

(ii) $\lim_{x \to a} (f + g)(x) = l + m$;

(iii) $\lim_{x \to a} (f \cdot g)(x) = lm$;

(iv) $\lim_{x \to a} (f/g)(x) = l/m$, provided $m \neq 0$.

Proof

(i) Let $\epsilon > 0$ be given. Then there exists $\delta > 0$ such that $|f(x) - l| < \epsilon$ for all x in dom $f \setminus \{a\}$ such that $|x - a| < \delta$. Hence, for all x in dom $f \setminus \{a\}$ such that $|x - a| < \delta$,

$$\big||f(x)| - |l|\big| \leq |f(x) - l| < \epsilon,$$

by Theorem 1.9, and so $\lim_{x \to a} |f|(x) = |l|$.

(ii) Let $\epsilon > 0$ be given. Then there exists $\delta_1 > 0$ such that $|f(x) - l| < \epsilon/2$ for all x in $\operatorname{dom} f \setminus \{a\}$ such that $|x - a| < \delta_1$, and there exists $\delta_2 > 0$ such that $|g(x) - m| < \epsilon/2$ for all x in $\operatorname{dom} g \setminus \{a\}$ such that $|x - a| < \delta_2$. Then, for all x in $\operatorname{dom}(f + g) \setminus \{a\}$ such that $|x - a| < \min \{\delta_1, \delta_2\}$,

$$|(f + g)(x) - (l + m)| = |(f(x) - l) + (g(x) - m)| \leq |f(x) - l| + |g(x) - m| < \epsilon.$$

(iii) If we take $\epsilon = 1$ then there exists $\delta_1 > 0$ such that $|f(x) - l| < 1$ for all x in $\operatorname{dom} f \setminus \{a\}$ such that $|x - a| < \delta_1$. Now,

$$|f(x)| - |l| \leq |f(x) - l|,$$

and so $|f(x)| < |l| + 1$ for all x in $\operatorname{dom} f \setminus \{a\}$ such that $|x - a| < \delta_1$. Let $\epsilon > 0$ be given. Provided $|x - a| < \delta_1$, we have that

$$\begin{aligned} |(f \cdot g)(x) - lm| &= |(f(x)g(x) - f(x)m) + (f(x)m - lm)| \qquad (3.12) \\ &\leq |f(x)| |g(x) - m| + |m| |f(x) - l| \\ &\leq (|l| + 1)|g(x) - m| + (|m| + 1) |f(x) - l| . \end{aligned}$$

Now, there exists $\delta_2 > 0$ such that $|f(x) - l| < \epsilon/(2(|m| + 1))$ for all x in $\operatorname{dom} f \setminus \{a\}$ such that $|x - a| < \delta_2$, and there exists $\delta_3 > 0$ such that $|g(x) - m| < \epsilon/(2(|l| + 1))$ for all x in $\operatorname{dom} g \setminus a$ such that $|x - a| < \delta_3$. From (3.12) it then follows that, if $0 < |x - a| < \min \{\delta_1, \delta_2, \delta_3\}$,

$$|(f \cdot g)(x) - lm| < (|l| + 1)\frac{\epsilon}{2(|l| + 1)} + |m|\frac{\epsilon}{2(|m| + 1)} = \epsilon.$$

(iv) We consider the function $1/g$. First, take $\epsilon = |m|/2 > 0$. Then there exists $\delta_1 > 0$ such that $|g(x) - m| < |m|/2$ for all x in $\operatorname{dom} g \setminus \{a\}$ such that $|x - a| < \delta_1$. Since

$$|m| - |g(x)| \leq |m - g(x)| < |m|/2,$$

we conclude that $|g(x)| > |m|/2$ for all x in $\operatorname{dom} g \setminus \{a\}$ such that $|x - a| < \delta_1$. Hence, provided $|x - a| < \delta_1$,

$$\left| \frac{1}{g(x)} - \frac{1}{m} \right| = \frac{|g(x) - m|}{|g(x)| |m|} < \frac{2|g(x) - m|}{|m|^2} . \qquad (3.13)$$

Now let $\epsilon > 0$ be given. There exists $\delta_2 > 0$ such that $|g(x) - m| < (|m|^2/2)\epsilon$ for all x in $\operatorname{dom} g \setminus \{a\}$ such that $|x - a| < \delta_2$. Hence, for all x such that $0 < |x - a| < \min \{\delta_1, \delta_2\}$, it follows from (3.13) that

$$\left| \frac{1}{g(x)} - \frac{1}{m} \right| < \frac{2|m|^2\epsilon}{2|m|^2} = \epsilon,$$

and so $\lim_{x \to a} (1/g(x)) = 1/m$. The statement concerning f/g $(= f \cdot (1/g))$ follows immediately from part (iii). $\qquad \square$

We sometimes want to distinguish between a **left limit** and a **right limit**. We say that l is the **left limit** of $f(x)$ as $x \to a$, and write $\lim_{x \to a-} f(x) = l$, if for every $\epsilon > 0$ there exists $\delta > 0$ such that $|f(x) - l| < \epsilon$ for all x in $(-\infty, a) \cap \operatorname{dom} f$ such that $|x - a| < \delta$. Similarly, l is the **right limit** of $f(x)$ as $x \to a$, or equivalently $\lim_{x \to a+} f(x) = l$, if for every $\epsilon > 0$ there exists $\delta > 0$ such that $|f(x) - l| < \epsilon$ for all x in $(a, \infty) \cap \operatorname{dom} f$ such that $|x - a| < \delta$.

The two limits may be different. For example, consider the function f given by $f(0) = 0$ and

$$f(x) = |x| + \frac{x}{|x|} \quad (x \in \mathbb{R}, \ x \neq 0), \tag{3.14}$$

with graph

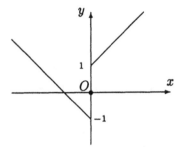

Then $\lim_{x \to 0-} f(x) = -1$, $\lim_{x \to 0+} f(x) = 1$.

Theorem 3.4

Let f be a real function. Then $\lim_{x \to a} f(x)$ exists if and only if $\lim_{x \to a-} f(x)$ and $\lim_{x \to a+} f(x)$ both exist and are equal. If

$$\lim_{x \to a-} f(x) = \lim_{x \to a+} f(x) = l$$

then $\lim_{x \to a} f(x) = l$.

Proof

Suppose first that $\lim_{x \to a} f(x) = l$. Then for all $\epsilon > 0$ there exists $\delta > 0$ such that $|f(x) - l| < \epsilon$ for all x in $\operatorname{dom} f \setminus \{a\}$ such that $|x - a| < \delta$. In particular, $|f(x) - l| < \epsilon$ for all x in $\operatorname{dom} f \cap (-\infty, a)$ such that $|x - a| < \delta$, and so $\lim_{x \to a-} f(x) = l$. Similarly, $\lim_{x \to a+} f(x) = l$.

Conversely, suppose that

$$\lim_{x \to a-} f(x) = \lim_{x \to a+} f(x) = l.$$

Let $\epsilon > 0$ be given. Then there exists $\delta_1 > 0$ such that $|f(x) - l| < \epsilon$ for all x in dom $f \cap (-\infty, a)$ such that $|x - a| < \delta_1$. Also, there exists $\delta_2 > 0$ such that $|f(x) - l| < \epsilon$ for all x in dom $f \cap (a, \infty)$ such that $|x - a| < \delta_2$. If we define $\delta = \min\{\delta_1, \delta_2\}$, we see that $|f(x) - l| < \epsilon$ for all x in dom $f \setminus \{a\}$ such that $|x - a| < \delta$, and so $\lim_{x \to a} f(x) = l$. $\qquad\square$

It is a consequence of this theorem that if the left and right limits at a are different, then the limit does not exist. Thus, for example,

$$\lim_{x \to 0} (|x| + (x/|x|))$$

does not exist.

We end this section by drawing attention to a crucially important limit concerning the function sin. A proper treatment of the circular functions will be given in Chapter 8, but at this stage we can at least make it plausible that

$$\lim_{\theta \to 0} \frac{\sin \theta}{\theta} = 1. \tag{3.15}$$

Consider the diagram in Fig. 3.7, in which C is the unit circle and P is a point

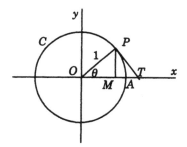

Figure 3.7. The limit of $\sin \theta / \theta$

on the circle. PT is the tangent to the circle at P and so is perpendicular to OP. Then

$$PT = PT/OP = \tan \theta, \quad MP = MP/OP = \sin \theta,$$

and (as long as we measure arc AP in radians) arc $AP = \theta$, with $0 < \theta < \pi/2$. It is then geometrically plausible that

$$MP < \text{arc } AP < PT,$$

and so $\sin \theta < \theta < \tan \theta$. Dividing by $\sin \theta$ gives

$$1 < \frac{\theta}{\sin \theta} < \frac{1}{\cos \theta},$$

and taking reciprocals gives

$$1 > \frac{\sin \theta}{\theta} > \cos \theta.$$

Since $\cos \theta \to 1$ as $\theta \to 0$, it follows that

$$\lim_{\theta \to 0+} \frac{\sin \theta}{\theta} = 1.$$

Since $\sin(-\theta) = -\sin \theta$ for all θ, we can conclude that

$$\lim_{\theta \to 0-} \frac{\sin \theta}{\theta} = \lim_{\theta \to 0+} \frac{\sin(-\theta)}{-\theta} = \lim_{\theta \to 0+} \frac{\sin \theta}{\theta} = 1.$$

Hence

$$\lim_{\theta \to 0} \frac{\sin \theta}{\theta} = 1. \tag{3.16}$$

Notice that the above analysis gives also the inequality

$$|\sin \theta| \le |\theta| \quad (\theta \in \mathbb{R}). \tag{3.17}$$

EXERCISES

3.14 Prove Theorem 3.1.

3.15 An alternative definition of the statement that $\lim_{x \to -\infty} f(x) = L$ is that for all $\epsilon > 0$ there exists $M > 0$ such that $|f(x) - L| < \epsilon$ for all $x < -M$. Show that this is equivalent to the definition in the text.

3.16 Show that an increasing function on a closed interval is bounded. Does this apply to an open interval?

3.17 Let A be a subset of $\mathbb{R}$, and let f be a bounded function with domain A. Show that, if $B \subseteq A$, then

$$\sup_{A} f \ge \sup_{B} f, \quad \inf_{A} f \le \inf_{B} f.$$

3.18 In Theorem 3.6(i) it is shown that if $\lim_{x \to a} f(x) = l$, then $\lim_{x \to a} |f(x)| = |l|$. Show by an example that the existence of $\lim_{x \to a} |f(x)|$ does not imply the existence of $\lim_{x \to a} f(x)$.

3.19 Prove that, if $\lim_{x \to a} f(x) = l$ and $\lim_{x \to a} g(x) = m$, then

$$\lim_{x \to a} \max \{f(x), g(x)\} = \max \{l, m\},$$
$$\lim_{x \to a} \min \{f(x), g(x)\} = \min \{l, m\}.$$

3.20 Show that $1 - \cos x = 2\sin^2(x/2)$, and deduce that

$$\lim_{x \to 0} \frac{1 - \cos x}{x^2} = \frac{1}{2}.$$

3.5 Continuity

Informally, a continuous function f is a real function with the property that a small change in x brings about a small change in $f(x)$. Such functions are important in applications, since most natural processes can be modelled with functions of this kind. There are exceptions, of course: a small increase in the electric potential within a storm cloud can have a very large effect; but such exceptions do not invalidate the importance of continuous functions within the total picture. The intuitive notion that f is continuous if its graph has no jumps is useful as far as it goes, but, as we have already seen, there are functions whose graphs cannot be drawn.

A formal definition is easily given: a real function f is **continuous at a point c in its domain** if $\lim_{x \to c} f(x) = f(c)$. That is to say, given $\epsilon > 0$ there exists $\delta > 0$ such that $|f(x) - f(c)| < \epsilon$ for all x in dom $f \setminus \{c\}$ such that $|x - c| < \delta$. The function f can fail to be continuous either because $\lim_{x \to c} f(x)$ does not exist (as with the function $x \mapsto |x| + (x/|x|)$ considered in (3.14)) or if $\lim_{x \to c} f(x) = l \ne f(c)$ (as with the function f defined by (3.9)).

Our first theorem regarding continuous functions provides a link with the theme of Chapter 2:

Theorem 3.5

Let f be a continuous function with domain $[a, b]$. Let (c_n) be a sequence with limit γ, such that c_n $(n \in \mathbb{N})$ and γ are in the interval $[a, b]$. Then the sequence $(f(c_n))$ has limit $f(\gamma)$.

Proof

From Theorem 3.3 we know that this holds for any polynomial function f, and indeed (provided we keep denominators clear of 0) for any rational function. We now see that it holds for any continuous function whatever. Let $\epsilon > 0$ be given. Then there exists $\delta > 0$ such that $|f(x) - f(\gamma)| < \epsilon$ for all x in $[a, b]$ such that $|x - \gamma| < \delta$. Also, there exists a positive integer N such that $|c_n - \gamma| < \delta$ whenever $n > N$. It follows that $|f(c_n) - f(\gamma)| < \epsilon$ whenever $n > N$, and hence $(f(c_n)) \to f(\gamma)$, as required. $\qquad\square$

We might decide to refer to the property established in this theorem as "sequential continuity", and remember the theorem in shorthand as "continuity implies sequential continuity". In fact we also have the converse implication:

Theorem 3.6

Let f be a function with domain $[a, b]$, and let $\alpha \in [a, b]$. Suppose that f has the property that $(f(c_n)) \to f(\gamma)$ for every sequence (c_n) lying entirely inside the interval $[a, b]$ with limit γ. Then f is continuous at γ.

Proof

Suppose, for a contradiction, that f is sequentially continuous but not continuous at α. Then, formally negating our standard definition of continuity, we conclude that there exists $\epsilon > 0$ with the property that, for all $\delta > 0$, there exists x in $(\gamma - \delta, \gamma + \delta)$ for which $|f(x) - f(\gamma)| \geq \epsilon$. We take $\delta = 1/n$ ($n = 1, 2, 3, \ldots$) and obtain elements x_n ($n = 1, 2, 3, \ldots$) with the property that $|x_n - \gamma| < 1/n$, but $|f(x_n) - f(\gamma)| \geq \epsilon$. It is clear that $(x_n) \to \gamma$, and that $(f(x_n)$ does not have limit $f(\gamma)$, and from this contradiction we deduce that f must after all be continuous at γ.   □

This result can be a useful device for showing that a function is *not* continuous at a given point. For example, consider again the function f given by

$$f(x) = \begin{cases} \sin \dfrac{1}{x} & \text{if } x \neq 0 \\ 0 & \text{if } x = 0. \end{cases}$$

Then the sequence $(2/(2n+1)\pi)$ has limit 0, but the sequence $(f(2/((2n+1)\pi)))$ is $(-1, 1, -1, 1, \ldots)$ and is certainly not convergent. It follows that f is not continuous at $x = 0$.

Example 3.7

Let $f : \mathbb{R} \to \mathbb{R}$ be given by

$$f(x) = \begin{cases} x \sin \dfrac{1}{x} & \text{if } x \neq 0 \\ 0 & \text{if } x = 0. \end{cases}$$

Show that f is continuous at 0.

Solution

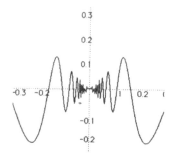

Figure 3.8. The function $x \mapsto x\sin(1/x)$

The graph of f is given in Fig. 3.8.

The function has zeros at $0, \pm(1/\pi), \pm(1/2\pi), \pm(1/3\pi), \ldots$. There are general theorems (to be proved later) to show that f is continuous at all points $a \neq 0$, but to establish the continuity at 0 we must go to the definition, and show that $\lim_{x \to 0} x\sin(1/x) = 0$. This is easily done. Let $\epsilon > 0$. Then

$$|x\sin(1/x) - 0| = |x\sin(1/x)| \leq |x|.$$

Hence, choosing $\delta = \epsilon$, we see that $|x\sin(1/x) - 0| < \epsilon$ for all $x \neq 0$ such that $|x - 0| < \delta$. $\qquad\square$

Our next function is much stranger:

Example 3.8

Let $f : \mathbb{R} \to \mathbb{R}$ be the function defined by

$$f(x) = \begin{cases} x & \text{if } x \text{ is rational} \\ 1 - x & \text{if } x \text{ is irrational.} \end{cases}$$

Show that f is continuous at the point a if and only if $a = 1/2$.

Solution

First, notice that $f(1/2) = 1/2$, and that

$$|f(x) - \tfrac{1}{2}| = \left\{ \begin{array}{ll} |x - \tfrac{1}{2}| & \text{if } x \text{ is rational} \\ |(1 - x) - \tfrac{1}{2}| & \text{if } x \text{ is irrational} \end{array} \right\} = |x - \tfrac{1}{2}|$$

for all x in $\mathbb{R}$. Hence, given $\epsilon > 0$ we may choose $\delta = \epsilon$ and obtain the required statement that $|f(x) - f(1/2)| < \epsilon$ for all $x \neq 1/2$ such that $|x - (1/2)| < \delta$.

On the other hand, suppose that $a \neq 1/2$, so that $a \neq 1 - a$, and suppose first that a is rational, so that $f(a) = a$. The sequence $(a + (\sqrt{2}/n))_{n \in \mathbb{N}}$ has limit a, while the sequence

$$(f(a + (\sqrt{2})/n))_{n \in \mathbb{N}} = (1 - a - (\sqrt{2})/n)_{n \in \mathbb{N}},$$

and has limit $1 - a$. It follows that f is not continuous at a.

So suppose now that a is irrational – which certainly implies that $a \neq 1 - a$. For each $n \geq 1$, let b_n be a rational number in the open interval $(a - \frac{1}{n}, a + \frac{1}{n})$. (This is possible by virtue of Exercise 1.7.) Then $|b_n - a| < \frac{1}{n}$, and so $(b_n) \to a$. On the other hand,

$$(f(b_n)) = (b_n) \to a \neq f(a),$$

since the definition gives $f(a) = 1 - a$. Thus f is not continuous at a. $\square$

Our next example, due to Dirichlet,[2] is even stranger.

Example 3.9

Let $f : [0, 1] \to \mathbb{R}$ be given by

$$f(x) = \begin{cases} 0 & \text{if } x \text{ is irrational} \\ 1/q & \text{if } x = p/q, \text{ with } p \in \mathbb{Z}, q \in \mathbb{N}, \text{ a fraction in lowest terms.} \end{cases}$$

Show that f is continuous at every irrational point and discontinuous at every rational point.

Solution

To show the first of these statements, we consider an irrational number a, and suppose that $\epsilon > 0$ is given. Let N be an integer such that $N \geq 1/\epsilon$. Observe now that the set $Q_N = \{p/q : 1 \leq q \leq N, 0 \leq p \leq q\}$, consisting of all rational numbers in $[0, 1]$ with denominator at most N, is finite. Since $a \notin Q_N$, it follows that δ, defined as $\min\{|a - x| : x \in Q_N\}$, is positive. Then all the rational numbers in the interval $(a - \delta, a + \delta)$ have denominator greater than N, and so, noting that $|f(x) - f(a)| = 0$ for every irrational x, we deduce that $|f(x) - f(a)| < 1/N \leq \epsilon$ for all x such that $|x - a| < \delta$. Thus f is continuous at a.

By contrast, let a be a rational number $\frac{p}{q}$, so that $f(x) = \frac{1}{q}$. The sequence $(\frac{p}{q} + \frac{\sqrt{2}}{n})$, has limit $\frac{p}{q}$, but the sequence $(f(\frac{p}{q} + \frac{\sqrt{2}}{n}))$ is the null sequence (0), with limit $0 \neq f(\frac{p}{q})$. Hence f is not continuous at a. $\square$

These are instructive examples, but analysis is not primarily about functions with strange and unexpected properties. Our main interest will be in

---

[2] Johann Peter Gustav Lejeune Dirichlet, 1805–1859

continuous functions, and at this stage we note that two very simple functions
are continuous at every real point.

Theorem 3.10

Let $C_k : \mathbb{R} \to \mathbb{R}$ and $i : \mathbb{R} \to \mathbb{R}$ (the constant and identity functions) be
defined, respectively, by (3.4). Then C_k and i are continuous at every point in
$\mathbb{R}$.

Proof

Let $\epsilon > 0$ be given, and let $a \in \mathbb{R}$. Then, for *every* $\delta > 0$ we can say, trivially,
that

$$|C_k(x) - C_k(a)| = 0 < \epsilon$$

for all $x \neq a$ such that $|x - a| < \delta$. The case of i is almost as easy: we choose
$\delta = \epsilon$ and observe that

$$|i(x) - i(a)| = |x - a| < \epsilon$$

for all $x \neq a$ such that $|x - a| < \delta$. $\square$

This essentially trivial theorem has far-reaching consequences, when combined with our next result:

Theorem 3.11

Let f and g be functions continuous at a. Then

(i) $|f|$ is continuous at a;

(ii) $f + g$ is continuous at a;

(iii) $f \cdot g$ is continuous at a;

(iv) $1/g$ is continuous at a, provided $g(a) \neq 0$.

Proof

This is a simple corollary of Theorem 3.3. From that theorem we deduce that

$$\lim_{x \to a} |f|(x) = |\lim_{x \to a} f(x)| = |f|(a),$$

and so $|f|$ is continuous at a. Also

$$\lim_{x \to a} (f + g)(x) = \lim_{x \to a} f(x) + \lim_{x \to a} g(x) = (f + g)(a),$$

and so $f + g$ is continuous at a. A similar argument applies to $f \cdot g$. As for $1/g$, from Theorem 3.3 we have that

$$\lim_{x \to a} (1/g)(x) = \frac{1}{\lim_{x \to a} g(x)} = (1/g)(a),$$

and so $1/g$ is continuous at a. $\square$

From our last two theorems we can deduce that every polynomial function is continuous at every point in $\mathbb{R}$, and that every rational function p/q, where p, q are polynomials, is continuous at every point in $\{x \in \mathbb{R} : q(x) \neq 0\}$.

We say that f is a **continuous function** if it is continuous at every point in its domain, and much interest will focus on functions f that are **continuous on a closed interval** $[a, b]$. The first important property of such a function is given by what is sometimes called the Intermediate Value Theorem, due to Bolzano.[3] If we think of the graph of a continuous function as having no jumps, then the theorem is intuitively clear. A proper proof, not surprisingly, makes use of the completeness property of $\mathbb{R}$, which can be thought of as telling us that there are no invisible gaps.

Theorem 3.12 (The Intermediate Value Theorem)

Let f be continuous on its domain $[a, b]$, and suppose that $f(a) < f(b)$. For each d such that $f(a) < d < f(b)$ there exists c in (a, b) such that $f(c) = d$.

Proof

Since f is continuous at a we may take $\epsilon = d - f(a)$ and assert that there exists $\delta > 0$ such that $|f(x) - f(a)| < d - f(a)$ for all x in $(a, b]$ such that $|x - a| < \delta$, that is to say, for all x in $(a, a + \delta)$. It follows that, for all x in $(a, a + \delta)$,

$$f(x) - f(a) \leq |f(x) - f(a)| < d - f(a)$$

and so $f(x) < d$. We deduce that the set

$$H = \{x \in (a, b) : f(x) < d\}$$

is non-empty. It is also bounded above (by b), and so by the Completeness Axiom (Property 1.2) has a supremum c.

We now show that $f(c) = d$, by showing that each of the alternatives, namely $f(c) < d$ and $f(c) > d$, leads to a contradiction. Suppose first that $f(c) < d$. Then we use the definition of continuity at c with $\epsilon = d - f(c)$, and

---
[3] Bernhard Placidus Johann Nepomuk Bolzano, 1781–1848

assert that there exists $\delta > 0$ with the property that $|f(x) - f(c)| < d - f(c)$ for all x in $[a, b] \setminus \{c\}$ such that $|x - c| < \delta$. We may safely assume that $\delta < b - c$. Now

$$f(x) - f(c) \leq |f(x) - f(c)| < d - f(c)\,,$$

and so $f(x) < d$ for all x in $[a, b] \setminus \{c\}$ such that $|x - c| < \delta$. So in particular we have $f(c + \frac{1}{2}\delta) < d$, which contradicts the definition of c as the least upper bound of H.

Suppose next that $f(c) > d$. We use the definition of continuity of f at c, with $\epsilon = d - f(c)$, to assert that there exists $\delta > 0$ with the property that $|f(x) - f(c)| < f(c) - d$ for all x in $[a, b] \setminus \{c\}$ such that $|x - c| < \delta$. Here we may safely assume that $\delta < c - a$. Now

$$f(c) - f(x) \leq |f(x) - f(c)| < f(c) - d\,,$$

and so $f(x) > d$ for all x in $[a, b] \setminus \{c\}$ such that $|x - c| < \delta$. Thus there is an interval $[c - \frac{1}{2}\delta, c]$ throughout which $f(x) > d$. It follows that $c - \frac{1}{2}\delta$ is an upper bound of H, a contradiction to the definition of c. $\qquad\square$

Remark 3.13

It is easy to modify the proof to deal with the case where $f(a) > f(b)$.

Example 3.14

Show that, for all integers $n \geq 2$, the function $x \mapsto x^{1/n}$ is continuous on $[0, \infty)$.

Solution

Let $x, a \in (0, \infty)$. We use the identity

$$p^n - q^n = (p - q)(p^{n-1} + p^{n-2}q + \cdots + pq^{n-2} + q^{n-1})\,.$$

Putting $p = x^{1/n}$ and $q = a^{1/n}$ gives

$$\begin{aligned}
|x - a| &= |x^{1/n} - a^{1/n}|\,|x^{(n-1)/n} + x^{(n-2)/n}a^{1/n} + \cdots + a^{(n-1)/n}| \\
&> |x^{1/n} - a^{1/n}|a^{(n-1)/n}
\end{aligned}$$

and so

$$|x^{1/n} - a^{1/n}| < \frac{|x - a|}{a^{(n-1)/n}}\,.$$

It follows that we can arrange for $|x^{1/n} - a^{1/n}|$ to be less than any given ϵ by taking $|x - a|$ less than $\epsilon a^{(n-1)/n}$. Thus the function is continuous at every a in $(0, \infty)$.

Continuity at 0 requires a separate argument, but is easily established, for we can make $x^{1/n}$ less than any given ϵ by choosing x less than ϵ^n. $\qquad\square$

Remark 3.15

The result of this example is in fact a corollary of a general result (Theorem 3.20) on inverse functions, to be established later. If n is odd the natural domain of the function $x \mapsto x^{1/n}$ is the whole of $\mathbb{R}$, and the function is continuous throughout its domain.

We look now at the circular functions sin, cos and tan, accepting that any truly rigorous statements about these functions must await a proper definition. Let $a \in \mathbb{R}$. From (3.7) we have that

$$|\sin x - \sin a| = \left| 2 \cos \frac{x+a}{2} \sin \frac{x-a}{2} \right| \leq 2 \left| \sin \frac{x-a}{2} \right|, \qquad (3.18)$$

since $\cos \theta \leq 1$ for all θ in $\mathbb{R}$. Now, from (3.17) we also have that $|\sin \theta| \leq |\theta|$ for all θ in $\mathbb{R}$, and so from (3.18) we obtain

$$|\sin x - \sin a| \leq |x - a|.$$

Thus, for every $\epsilon > 0$ there exists $\delta > 0$ (namely $\delta = \epsilon$) such that $|\sin x - \sin a| < \epsilon$ for all $x \neq a$ such that $|x - a| < \delta$. Thus sin is continuous at a.

The proof that cos is continuous proceeds in exactly the same way, beginning with the observation (again see (3.7)) that

$$|\cos x - \cos a| = \left| 2 \sin \frac{x+a}{2} \sin \frac{x-a}{2} \right| \leq 2 \left| \sin \frac{x-a}{2} \right|.$$

From Theorem 3.11, the function tan $= \sin / \cos$ is continuous except where cos takes the value 0, that is to say, except at $x = (2n+1)\pi/2, \quad n \in \mathbb{Z}$.

We end this section with a general theorem concerning compositions of functions:

Theorem 3.16

Let f g be continuous functions such that im $f \subseteq$ dom g. Then $g \circ f$, with domain dom f, is continuous.

Proof

Let $a \in$ dom f, and let $b = f(a)$ ($\in$ dom g). Let $\epsilon > 0$ be given. Then there exists $\delta' > 0$ such that $|g(y) - g(b)| < \epsilon$ for all y in dom $g \setminus \{b\}$ for which $|y - b| < \delta'$. Also, there exists $\delta > 0$ such that $|f(x) - f(a)| < \delta'$ for all x in dom $f \setminus \{a\}$ for which $|x - a| < \delta$. It follows that, for all x in dom$(g \circ f) \setminus \{a\}$ for which $|x - a| < \delta$,

$$|(g \circ f)(x) - (g \circ f)(a)| = |g(f(x)) - g(f(a))| < \epsilon.$$

Thus $g \circ f$ is continuous at a.   □

We remark that the theorems now established show that quite complicated functions such as $x \mapsto \sin(x^3 + 1)$, $x \mapsto 3\cos(x^2)/(\sin^2 x + 4)$ are continuous for all x. Also, $x \mapsto x\sin(1/x)$, which we showed to be continuous at $x = 0$, is in fact continuous for all x.

EXERCISES

3.21 Let f, g be continuous on $[a, b]$. Show that $\max\{f, g\}$ and $\min\{f, g\}$ are also continuous on $[a, b]$.

3.22 Let $a, b \in \mathbb{R}$, with $a < b$. Discuss the continuity of

$$\sqrt{(x - a)(b - x)}, \quad \sqrt{\frac{x - a}{b - x}}.$$

3.23 For which values of x is $\cot x$ continuous?

3.24 Let f be continuous on $[a, b]$, and suppose that $f(c) \neq 0$ for some c in (a, b). Show that there exists $\delta > 0$ with the property that $f(x) \neq 0$ for every x in $(c - \delta, c + \delta)$.

3.25 Let $f : [0, 1] \to \mathbb{R}$ be given by

$$f(x) = \begin{cases} x & \text{if } x \text{ is rational} \\ x^2 & \text{if } x \text{ is irrational.} \end{cases}$$

Show that f is continuous at 0 and at 1, but is not continuous at any point in $(0, 1)$.

3.26 Let f be a continuous function with domain and image $[a, b]$. Show that there exists c in $[a, b]$ such that $f(c) = c$.

3.27 Let f and g be continuous on $[0, 1]$, and suppose that $f(0) < g(0)$, $f(1) > g(1)$. Show that there exists x in $(0, 1)$ such that $f(x) = g(x)$. Deduce that the equation

$$\frac{x + 1}{3} = \sin\frac{\pi x}{2}$$

has a solution in $(0, 1)$.

3.28 Let f be continuous on $[a, b]$ and suppose that it takes every real value at most once. Show that f is monotonic.

3.29 Let $f : (-1, 1) \to \mathbb{R}$ be continuous at 0, and suppose that $f(x) = f(x^2)$ for all x in $(-1, 1)$. Show that $f(x) = f(0)$ for all x in $(-1, 1)$.

3.30 Let $a, b > 1$, and let f be a bounded function on $[0, 1]$ such that

$$f(ax) = bf(x) \quad (0 \le x \le 1/a).$$

Show that f is continuous at 0.

3.31 Deduce Theorem 3.11 from Theorem 2.8, using Theorems 3.5 and 3.6.

3.32 Prove Theorem 3.16 using Theorems 3.5 and 3.6.

3.6 Uniform Continuity

It is sometimes useful to know that a function on a closed interval $[a, b]$ is not just continuous but **uniformly continuous**, that is to say, has the property that, for all $\epsilon > 0$ there exists $\delta > 0$ such that, for all x, y in $[a, b]$,

$$|x - y| < \delta \implies |f(x) - f(y)| < \epsilon. \tag{3.19}$$

That is, the *same* δ works in the definition throughout the interval $[a, b]$.

To see the difference between continuity and uniform continuity, consider the function $x \mapsto 1/x$ defined on the open interval $(0, 1)$. This is continuous throughout its domain, but is not uniformly continuous. Suppose, for a contradiction, that for each $\epsilon > 0$ there exists $\delta > 0$ such that (3.19) holds. For all x, y in $(0, 1)$ we have

$$|f(x) - f(y)| = \left| \frac{1}{x} - \frac{1}{y} \right| = \frac{|x - y|}{xy} < \epsilon$$

if and only if $|x - y| < xy\epsilon$. That is, we require that our fixed δ has the property that $\delta \le xy\epsilon$ for **all** x, y in $(0, 1)$. This is not possible since, however small δ may be, we can always choose x and y small enough to make $xy\epsilon$ less than δ.

Intuitively, what is happening here is that the steepness of the function determines at each point the δ we require in order to demonstrate continuity. As x and y approach 0 the steepness increases, and we require smaller and smaller values of δ. There is no single δ that will do for the whole interval $(0, 1)$.

It is not accidental that in the example above the domain is an open interval, for we have the following result:

Theorem 3.17

Let f be a continuous function on the closed interval $[a, b]$. Then f is uniformly continuous on $[a, b]$.

This is quite a deep result, and we approach the proof in a way that seems oblique. An **open covering** of $[a, b]$ is a possibly infinite collection $\mathcal{C}$ of open intervals I_j ($j \in J$) with the property that every x in $[a, b]$ belongs to at least one interval I_j. We now establish what is always called the Heine[4]–Borel[5] Theorem:

Theorem 3.18

Let $[a, b]$ be a closed interval, and let $\mathcal{C}$ be an open covering of $[a, b]$, by open intervals I_j ($j \in J$). Then there exists a finite collection $\{I_{j_1}, \ldots, I_{j_m}\}$ of open intervals in $\mathcal{C}$ such that

$$[a, b] \subseteq I_{j_1} \cup \ldots \cup I_{j_m} .$$

Proof

Let us suppose, by way of contradiction, that $[a, b]$ cannot be covered by the intervals from a finite subcollection of $\mathcal{C}$. Then at least one of the two halves $[a, \frac{1}{2}(a + b)]$, $[\frac{1}{2}(a + b), b]$ of $[a, b]$ cannot be covered by a finite subcollection from $\mathcal{C}$. So there is an interval $[a_1, b_1]$ contained in $[a, b]$ and of length $(b - a)/2$ that cannot be covered by a finite subcollection from $\mathcal{C}$. We can repeat this bisection process indefinitely, obtaining, for $n = 1, 2, \ldots$, an interval $[a_n, b_n]$ of length $(b - a)/2^n$ that cannot be covered by a finite subcollection from $\mathcal{C}$. Moreover, we have

$$[a, b] \supset [a_1, b_1] \supset [a_2, b_2] \supset \cdots,$$

and from this it follows that (a_n) is a monotonic increasing sequence bounded above by b, while (b_n) is a monotonic decreasing sequence, bounded below by a. By Theorems 2.9 and 2.10, both sequences converge: let

$$\alpha = \lim_{n \to \infty} a_n, \quad \beta = \lim_{n \to \infty} b_n .$$

Then

$$\beta - \alpha = \lim_{n \to \infty} (b_n - a_n) = \lim_{n \to \infty} \frac{b - a}{2^n} = 0 ,$$

and so $\alpha = \beta$. Observe at this point that $\alpha \in [a_n, b_n]$ for *every* n.

---

[4] Heinrich Eduard Heine, 1821–1881
[5] Félix Édouard Justin Émile Borel, 1871–1956

Now $\alpha \in I_j = (c_j, d_j)$ for some open interval I_j in the open covering $\mathcal{C}$. If we choose N large enough so that $b_N - a_N < \min\{\alpha - c_i, d_j - \alpha\}$, then, since $\alpha \in [a_N, b_N]$, we must have $[a_N, b_N] \subseteq (c_j, d_j) = I_j$.

$$\overline{\quad\quad\quad\underset{c_j}{\bullet}\quad\quad\underset{a_N}{\bullet}\quad\underset{\alpha}{\bullet}\quad\quad\underset{b_N}{\bullet}\ \underset{d_j}{\bullet}\quad\quad\quad}$$

Thus the interval $[a_N, b_N]$, which by construction was not covered by any finite subcollection of intervals from $\mathcal{C}$, is in fact covered by the single interval I_j. This contradiction completes the proof. $\qquad\qquad\qquad\qquad\qquad\qquad\qquad\square$

We can now prove Theorem 3.17. Let $\epsilon > 0$ be given. Since f is by assumption continuous throughout $[a, b]$, at each point c in $[a, b]$ we can find $\delta_c > 0$, depending on c, such that

$$|f(x) - f(c)| < \epsilon/2 \text{ for all } x \text{ in } [a, b] \cap (c - \delta_c, c + \delta_c).$$

The collection

$$\mathcal{C} = \{(c - \tfrac{1}{2}\delta_c, c + \tfrac{1}{2}\delta_c) \, : \, c \in [a, b]\}$$

is certainly an open covering of $[a, b]$, and so, by Theorem 3.18, a finite subcollection

$$\{(c_1 - \tfrac{1}{2}\delta_1, c_1 + \tfrac{1}{2}\delta_1), \dots, (c_m - \tfrac{1}{2}\delta_m, c_m + \tfrac{1}{2}\delta_m)\}$$

also covers $[a, b]$. (We are simplifying the notation slightly, writing δ_i rather than δ_{c_i}.)

For $k = 1, \dots, m$, we have that $|f(x) - f(c_k)| < \epsilon/2$ for all x in $[a, b] \cap (c_k - \delta_k, c_k + \delta_k)$. Let

$$\delta = \min\{\tfrac{1}{2}\delta_1, \dots, \tfrac{1}{2}\delta_m\},$$

and suppose that x, y in $[a, b]$ are such that $|x - y| < \delta$. There exists k such that $x \in (c_k - \tfrac{1}{2}\delta_k, c_k + \tfrac{1}{2}\delta_k)$. Then

$$|y - c_k| = |(y - x) + (x - c_k)| \leq |y - x| + |x - c_k|$$
$$< \delta + \tfrac{1}{2}\delta_k \leq \tfrac{1}{2}\delta_k + \tfrac{1}{2}\delta_k = \delta_k$$

and so $|f(y) - f(c_k)| < \epsilon/2$. It follows that

$$|f(x) - f(y)| = |(f(x) - f(c_k)) + (f(c_k) - f(y))|$$
$$\leq |f(x) - f(c_k)| + |f(y) - f(c_k)| < \epsilon.$$

This completes the proof of Theorem 3.17. $\qquad\qquad\qquad\qquad\qquad\qquad\square$

A further consequence of the Heine–Borel Theorem is the following important result:

Theorem 3.19

Let f be continuous on $[a, b]$. Then f is bounded on $[a, b]$, and attains both its supremum and infimum. That is, there exist c, d in $[a, b]$ such that $f(c) = \sup_{[a,b]} f$, $f(d) = \inf_{[a,b]} f$.

Proof

Let $\epsilon = 1$. By continuity we can assert that for each c in $[a, b]$ there exists $\delta_c > 0$ (depending on c) such that $|f(x) - f(c)| < 1$ for all x in $[a, b]$ such that $|x - c| < \delta$. The set

$$\mathcal{C} = \{(c - \delta_c, c + \delta_c) : c \in [a, b]\}$$

is an open covering, and so, by Theorem 3.18, there is a finite subcollection

$$\{(c_1 - \delta_1, c_1 + \delta_1), \ldots, (c_m - \delta_m, c_m + \delta_m)\}$$

covering $[a, b]$. Let

$$K = \max \{f(c_1) + 1, \ldots, f(c_m) + 1\} \quad k = \min \{f(c_1) - 1, \ldots, f(c_m) - 1\}.$$

Let $x \in [a, b]$. Then there exists at least one i in $\{1, \ldots, m\}$ such that $x \in (c_i - \delta_i, c + \delta_i)$. Hence

$$k \le f(c_i) - 1 < f(x) < f(c_i) + 1 \le K,$$

and so f is bounded on $[a, b]$.

To establish the second part of the theorem, let $M = \sup_{[a,b]} f$ and $m = \inf_{[a,b]} f$, and suppose, for a contradiction, that there does *not* exist c such that $f(c) = M$. Then the function $x \mapsto M - f(x)$ takes positive values throughout $[a, b]$, and so the function

$$x \mapsto 1/(M - f(x)) \quad (x \in [a, b]) \tag{3.20}$$

is continuous throughout $[a, b]$, by Theorem 3.11. On the other hand, we know that, for every positive P, there exists x in $[a, b]$ such that $0 < M - f(x) < 1/P$, since otherwise we would have an upper bound $M - (1/P)$ for f that is smaller than M. Hence, for all $P > 0$ there exists x in $[a, b]$ such that $1/(M - f(x)) > P$. Thus the function given by (3.20) is continuous but not bounded in $[a, b]$, and this contradiction establishes that $\sup_{[a,b]} f$ is attained.

Similarly, by considering the function $x \mapsto 1/(f(x) - m)$, we establish that $\inf_{[a,b]} f$ is attained. $\square$

It is important to note that it is necessary to have a *closed* interval in the above theorem. In the interval $(0, 1]$, as we have seen, the function $x \mapsto 1/x$ is continuous, but not bounded. It *is* bounded on any closed interval, such as $[0.001, 1]$, contained in $(0, 1]$.

EXERCISES

3.33 Let $a, b, c \in \mathbb{R}$, with $a < b < c$. Show that if f is uniformly continuous on $[a, b]$ and also on $[b, c]$, then it is uniformly continuous on $[a, c]$.

3.7 Inverse Functions

We shall often want to use the notation $C[a, b]$ for the set of all real functions f whose domain is the closed interval $[a, b]$ and which are continuous at all points in $[a, b]$.

In this section we shall consider a strictly increasing function f in $C[a, b]$. For such a function we immediately conclude that $f(x) \in [f(a), f(b)]$ for every x in $[a, b]$. Indeed we have im $f = [f(a), f(b)]$ by the Intermediate Value Theorem (Theorem 3.12), since for every d in $[f(a), f(b)]$ there exists c in $[a, b]$ such that $f(c) = d$. The strictly increasing property then ensures that c is unique, for if we have c_1, c_2 with (say) $c_1 < c_2$ and $f(c_1) = f(c_2) = d$, then we have an immediate contradiction to the increasing property. Consequently we have an **inverse function** $f^{-1} : [f(a), f(b)] \to [a, b]$ defined by the rule that, *for each y in $[f(a), f(b)]$, $f^{-1}(y)$ is the unique x in $[a, b]$ such that $f(x) = y$.*

It is not hard to see that f^{-1} is strictly increasing. Let t, u in $[f(a), f(b)]$ be such that $t < u$, and write $f^{-1}(t) = x$, $f^{-1}(u) = y$. If $x = y$ then $t = f(x) = f(y) = u$, a contradiction. If $y < x$ then $u = f(y) < f(x) = t$, again a contradiction. The only remaining possibility is that

$$f^{-1}(t) = x < y = f^{-1}(u),$$

and so f^{-1} is increasing.

We have the following theorem:

Theorem 3.20

Let f be continuous on $[a, b]$, and suppose that f is increasing. Then there exists an inverse function $f^{-1} : [f(a), f(b)] \to [a, b]$. The function f^{-1} is strictly increasing and continuous.

Proof

It remains to prove that f^{-1} is continuous at all points in $[f(a), f(b)]$. Let $d \in [f(a), f(b)]$, and let $c = f^{-1}(d)$. Let $\epsilon > 0$ be given. We may certainly assume that $\epsilon \leq \min \{c - a, b - c\}$, so that $c + \epsilon, c - \epsilon \in [a, b]$. Now $f(c + \epsilon) > f(c) = d$, and so we may write $f(c+\epsilon) = d + \delta_1$, with $\delta_1 > 0$. Similarly, we may write $f(c-\epsilon) = d - \delta_2$, with $\delta_2 > 0$. If $\delta = \min\{\delta_1, \delta_2\}$, then, for all y in $(d-\delta, d+\delta)$,

$$f^{-1}(y) < f^{-1}(d+\delta_1) = c + \epsilon, \quad f^{-1}(y) > f^{-1}(d - \delta_2) = c - \epsilon.$$

That is, $|f^{-1}(y) - f^{-1}(d)| < \epsilon$ for all y in $[f(a), f(b)]$ such that $|y - d| < \delta$, exactly as required. $\qquad\square$

An argument very similar to that given above shows that:

Theorem 3.21

Let f be continuous on $[a, b]$, and suppose that f is strictly decreasing. Then there exists an inverse function $f^{-1} : [f(b), f(a)] \to [a, b]$. The function f^{-1} is strictly decreasing and continuous.

One of many consequences of Theorem 3.20 is that the function $x \mapsto \sqrt{x}$, the inverse of the continuous function $x \mapsto x^2$, is strictly increasing and continuous in any closed interval contained in $[0, \infty)$, and thus in $[0, \infty)$ itself. More generally, we can assert that the function $x \mapsto x^{1/n}$ is a continuous strictly increasing function on $\mathbb{R}$ for every odd natural number n, and is a continuous increasing function on $[0, \infty)$ for every even natural number n.

Also, since the function $\sin \in C[-\pi/2, \pi/2]$ and is strictly increasing in that interval, with image $[-1, 1]$, there is an inverse function $\sin^{-1} : [-1, 1] \to [-\pi/2, \pi/2]$ which is also strictly increasing and continuous.

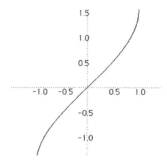

Figure 3.9. The function $\sin^{-1}$

By contrast, the continuous function $\cos : [0, \pi] \to [-1, 1]$ is strictly decreasing, and so the inverse function $\cos^{-1} : [-1, 1] \to [0, \pi]$ is continuous and strictly decreasing.

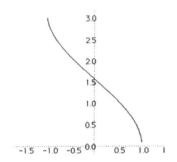

Figure 3.10. The function $\cos^{-1}$

The function $\tan : (-\pi/2, \pi/2) \to \mathbb{R}$ is strictly increasing and continuous, and so there is a strictly increasing continuous function $\tan^{-1} : \mathbb{R} \to (-\pi/2, \pi/2)$.

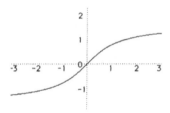

Figure 3.11. The function $\tan^{-1}$

These examples illustrate the fact that we can often obtain an inverse for a function by restricting the domain. We do have to be careful: certainly $\sin(\sin^{-1} x) = x$ for all x in $[-1, 1]$, but $\sin^{-1}(\sin x) = x$ only if $x \in [-\pi/2, \pi/2]$.

An example concerning quadratic functions illustrates the main points:

Example 3.22

Consider the function $f : \mathbb{R} \to \mathbb{R}$ defined by

$$f(x) = x^2 + 2x - 3 \quad (x \in \mathbb{R}).$$

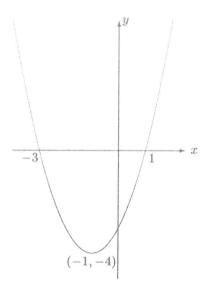

Since $x^2 + 2x - 3 = (x + 1)^2 - 4$, the function takes its minimum value of -4 when $x = -1$. The zeros of the function are $x = -3$ and $x = 1$. If we restrict the domain to the interval $[-1, \infty)$, we obtain a strictly increasing function whose image is $[-4, \infty)$. The formula for the inverse function is obtained by solving the equation $y = x^2 + 2x - 3$ for x in terms of y:

$$y = -1 \pm \sqrt{y + 4}.$$

We want the image of f^{-1} to be $[-1, \infty)$, so we choose the positive sign:

$$f^{-1}(y) = -1 + \sqrt{y + 4} \quad (y \in [-4, \infty)).$$

If, alternatively, we restrict the domain of f to $(-\infty, -1]$ we have a strictly decreasing function, whose inverse is the strictly decreasing function $y \mapsto -1 - \sqrt{y + 4} \quad (y \in [-4, \infty))$.

EXERCISES

3.34 Show that, for all x in $[-1, 1]$,

$$\sin^{-1} x + \cos^{-1} x = \frac{\pi}{2}.$$

3.35 Show that the "weak" increasing property

$$x \le y \implies f(x) \le f(y)$$

is not sufficient for there to be an inverse function.

3.36 In the spirit of Example 3.22, find an inverse function, with suitably restricted domain, for the function $x \mapsto 8 + 2x - x^2$.

3.37 Determine the inverse of the function $f : \mathbb{R} \setminus \{1\} \to \mathbb{R}$ given by

$$f(x) = \frac{1}{1-x} \, .$$

What is its domain?

4
Differentiation

4.1 The Derivative

The idea of a derivative, going back to Newton and Leibniz in the seventeenth century, is central to the application of mathematics to changing situations. It was primarily motivated by physics, but the ideas are relevant to a wide variety of disciplines.

Let f be a real function, and suppose that c is a point in the **interior** of dom f, by which we mean that there is an open interval $(c-\eta, c+\eta)$ of c wholly contained in dom f. If, as is normally the case, dom f is a closed interval $[a, b]$, the interior of dom f is the open interval (a, b). If

$$\lim_{x \to c} \frac{f(x) - f(c)}{x - c} \tag{4.1}$$

exists, we say that f **is differentiable at** c. In that case we denote the limit (4.1) by $f'(c)$, and say that $f'(c)$ is the **derivative** (or the **differential coefficient**) of f at c. The process of moving from f to f' is called **differentiation**.

We can let c vary over the interior of dom f, and think of f' as another function, "derived" from f. Other notations are used for $f'(x)$, such as

$$D_x f(x), \quad \frac{d}{dx} f(x), \quad \frac{df}{dx}. \tag{4.2}$$

An alternative version of (4.1) gives

$$f'(c) = \lim_{h \to 0} \frac{f(c + h) - f(c)}{h}. \tag{4.3}$$

The first and simplest example concerns the function $f : x \mapsto x^n$, where n is a positive integer. The algebraic identity

$$x^n - c^n = (x - c)(x^{n-1} + cx^{n-2} + c^2 x^{n-3} + \cdots + c^{n-2}x + c^{n-1})$$

is easily verified, and it follows that

$$f'(c) = \lim_{x \to c}(x^{n-1} + cx^{n-2} + c^2 x^{n-3} + \cdots + c^{n-2}x + c^{n-1}) = nc^{n-1}. \quad (4.4)$$

The function $x \mapsto x^n$ is differentiable at every real x, and $f'(x) = nx^{n-1}$. In fact, we shall see that all polynomial functions are differentiable.

First, however, it important to establish that differentiability is a stronger property than continuity:

Theorem 4.1

Let f be a real function, differentiable at c. Then f is continuous at c.

Proof

Since

$$\lim_{x \to c} \frac{f(x) - f(c)}{x - c} = f'(c),$$

and $\lim_{x \to c}(x - c) = 0$, it follows from Theorem 3.3 that

$$\lim_{x \to c}(f(x) - f(c)) = \lim_{x \to c}(x - c)f'(c) = 0.$$

Thus $\lim_{x \to c} f(x) = f(c)$, and so f is continuous at c. $\qquad \square$

The converse need not hold. For example, the function $x \mapsto |x|$ is continuous at $x = 0$, but it is not differentiable, since the left and right limits of $(f(x) - f(0))/(x - 0)$ are different:

$$\lim_{x \to 0+} \frac{f(x) - f(0)}{x - 0} = \lim_{x \to 0+} \frac{x}{x} = 1, \quad \lim_{x \to 0-} \frac{f(x) - f(0)}{x - 0} = \lim_{x \to 0+} \frac{-x}{x} = -1.$$

The intuitive idea that a function is continuous if its graph has no gaps, and differentiable if it has no corners, is fine as far as it goes, but is not enough to deal with the function $x \mapsto x \sin(1/x)$. (See Exercise 4.3.) In Chapter 9 we shall encounter a strange function which is continuous throughout its domain but is differentiable nowhere.

The next theorem is a cornerstone of what is called "differential calculus", a set of rules and procedures for finding derivatives:

Theorem 4.2

Let f, g be real functions, differentiable at c.

(i) kf is differentiable at c for every constant k, and $(kf)' = kf'$.

(ii) $f + g$ is differentiable at c, and $(f + g)' = f' + g'$.

(iii) $f \cdot g$ is differentiable at c, and $(f \cdot g)' = f \cdot g' + f' \cdot g$.

(iv) $1/g$ is differentiable at c provided $g(c) \neq 0$, and $(1/g)' = -g'/g^2$.

Proof

(i) We have

$$\lim_{x \to c} \frac{(kf)(x) - (kf)(c)}{x - c} = \lim_{x \to c} k \cdot \left[\frac{f(x) - f(c)}{x - c} \right].$$

Hence, by Theorem 3.3, the limit exists, and is equal to $kf'(c)$.

(ii) We have

$$\lim_{x \to c} \frac{(f + g)(x) - (f + g)(c)}{x - c} = \lim_{x \to c} \left[\frac{f(x) - f(c)}{x - c} + \frac{g(x) - g(c)}{x - c} \right].$$

Hence, by Theorem 3.3, the limit exists, and is equal to $f'(c) + g'(c)$.

(iii) Here we have

$$\lim_{x \to c} \frac{(f \cdot g)(x) - (f \cdot g)(c)}{x - c}$$

$$= \lim_{x \to c} \frac{f(x)g(x) - f(x)g(c) + f(x)g(c) - f(c)g(c)}{x - c}$$

$$= \lim_{x \to c} \left[f(x) \cdot \frac{g(x) - g(c)}{x - c} + \frac{f(x) - f(c)}{x - c} \cdot g(c) \right].$$

Hence, by the continuity of f and by Theorem 3.3, the limit exists, and is equal to $f(c)g'(c) + f'(c)g(c)$.

(iv) We note first that the continuity of g implies that there is an interval $(c - \delta, c + \delta)$ within which $g(x) \neq 0$. Then

$$\lim_{x \to c} \frac{(1/g)(x) - (1/g)(c)}{x - c} = \lim_{x \to c} \frac{1}{x - c} \left(\frac{1}{g(x)} - \frac{1}{g(c)} \right)$$

$$= \lim_{x \to c} -\frac{1}{x - c} \cdot \frac{g(x) - g(c)}{g(x)g(c)}$$

$$= \lim_{x \to c} -\frac{g(x) - g(c)}{x - c} \cdot \frac{1}{g(x)g(c)}.$$

Since g is continuous at c and $g(c) \neq 0$, it follows by Theorem 3.3 that the limit exists, and is equal to $-g'(c)/[g(c)]^2$. □

Corollary 4.3

Let f and g be real functions, differentiable at c. Then

(i) $f - g$ is differentiable at c, and $(f - g)' = f' - g'$;

(ii) f/g is differentiable at c provided $g(c) \neq 0$, and

$$(f/g)' = (f' \cdot g - f \cdot g')/g^2.$$

Proof

(i) This is immediate, since $f - g = f + (-1)g$.

(ii) It is clear from (iii) and (iv) above that $f/g = f \cdot (1/g)$ is differentiable at c. Also, by (iii) and (iv),

$$(f/g)' = f \cdot (1/g)' + f' \cdot (1/g) = -\frac{f \cdot g'}{g^2} + \frac{f'}{g} = \frac{f' \cdot g - f \cdot g'}{g^2},$$

as required.   □

As a consequence of the theorem and its corollary, every polynomial function is differentiable at every real x, and every rational function p/q (where p and q are polynomials) is differentiable at every point in $\{x \in \mathbb{R} : q(x) \neq 0\}$.

The question of composition of functions is a little more difficult:

Theorem 4.4

Let f and g be real functions, and suppose that f is differentiable at c and g is differentiable at $f(c)$. Then $g \circ f$ is differentiable at c, and $(g \circ f)' = (g' \circ f) \cdot f'$.

Proof

On the face of it, we could argue as follows:

$$\frac{g(f(x)) - g(f(c))}{x - c} = \frac{g(f(x)) - g(f(c))}{f(x) - f(c)} \cdot \frac{f(x) - f(c)}{x - c}.$$

Hence, if we write $f(x)$ as u and $f(c)$ as d, we have

$$\frac{g(f(x)) - g(f(c))}{x - c} = \frac{g(u) - g(d)}{u - d} \cdot \frac{f(x) - f(c)}{x - c}.$$

If we let $x \to c$, then $u \to d$ by the continuity of f, and so

$$(g \circ f)'(c) = g'(d)f'(c) = [(g' \circ f) \cdot f'](c),$$

as required. The flaw in this argument is that $f(x) - f(c)$ may well be zero, and could be zero infinitely often in any interval containing c. So we need a more complicated argument that takes account of this possibility.

Write $f(c) = d$, and define a new function h by the rule that, for all u in $\operatorname{im} f$,

$$h(u) = \begin{cases} \dfrac{g(u) - g(d)}{u - d} - g'(d) & \text{if } u \neq d \\[2mm] 0 & \text{if } u = d. \end{cases}$$

Since $g'(d)$ exists, it is clear that $\lim_{u \to d} h(u) = 0$. Thus h is continuous at d, and so, by Theorem 3.16, $h \circ f$ is continuous at c. From the equality

$$(h \circ f)(x) = \frac{g(f(x)) - g(d)}{f(x) - d} - g'(d) \quad (f(x) \neq d)$$

we deduce that

$$g(f(x)) - g(d) = [(h \circ f)(x) + g'(d)](f(x) - d),$$

and this equality is trivially valid even if $f(x) = d$.

Suppose now that $x \neq c$, and divide this last equality by $x - c$ to obtain

$$\frac{g(f(x)) - g(f(c))}{x - c} = [(h \circ f)(x) + g'(d)] \cdot \frac{f(x) - f(c)}{x - c}.$$

Taking the limit as $x \to c$, and noting that $\lim_{x \to c}(h \circ f)(x) = h(f(c)) = 0$, we obtain

$$(g \circ f)'(c) = g'(f(c))f'(c) = [(g' \circ f) \cdot f'](c),$$

exactly as required. $\qquad \square$

We remark that the usual "calculus" way of remembering the "chain rule" is very helpful: if we write $f(x)$ as u and $g(x)$ as y, the rule states that

$$\frac{dy}{dx} = \frac{dy}{du} \cdot \frac{du}{dx}.$$

Thus, for example, the function $x \to (x^3 + x^2 + 3)^6$, is expressible as $g \circ f$, where $f(x) = x^3 + x^2 + 3$ and $g(u) = u^6$. We have $g'(u) = 6u^5$ and $f'(x) = 3x^2 + 2x$, and so, by the theorem

$$(g \circ f)'(x) = g'(f(x)) f'(x) = 6(x^3 + x^2 + 3)^5(3x^2 + 2x).$$

In other words, we have

$$\frac{dy}{du} = 6u^5, \quad \frac{du}{dx} = 3x^2 + 2x,$$

and so

$$\frac{dy}{dx} = \frac{dy}{du} \cdot \frac{du}{dx} = 6(x^3 + x^2 + 3)^5(3x^2 + 2x).$$

We end this section by considering the differentiability of the functions sin and cos:

Theorem 4.5

Let $f(x) = \sin x$, $g(x) = \cos x$. Then

$$f'(x) = \cos x, \quad g'(x) = -\sin x.$$

Proof

Let $c \in \mathbb{R}$. Then

$$\frac{\sin x - \sin c}{x - c} = \left(2 \cos \frac{x+c}{2} \sin \frac{x-c}{2}\right) \Big/ (x - c), \text{ by (3.7)}$$

$$= \cos \frac{x+c}{2} \cdot \left(\sin \frac{x-c}{2} \Big/ \left(\frac{x-c}{2}\right)\right) \to \cos c \text{ as } x \to c,$$

by (3.15) and since cos is continuous. Thus $f'(c) = \cos c$.
    Similarly,

$$\frac{\cos x - \cos c}{x - c} = = \left(-2 \sin \frac{x+c}{2} \sin \frac{x-c}{2}\right) \Big/ (x - c), \text{ by (3.7)}$$

$$= -\sin \frac{x+c}{2} \cdot \left(\sin \frac{x-c}{2} \Big/ \left(\frac{x-c}{2}\right)\right) \to -\sin c \text{ as } x \to c,$$

by (3.15) and since sin is continuous. Thus $f'(c) = -\sin c$ $\qquad\square$

EXERCISES

4.1 Show that the function $f : x \mapsto \sqrt{|x|}$ is continuous but not differentiable at $x = 0$.

4.2 Consider a function $f : [a, b] \to \mathbb{R}$. For $c \in (a, b]$ the **left derivative** $f'_l(c)$ is defined by

$$f'_l(c) = \lim_{x \to c-} \frac{f(x) - f(c)}{x - c},$$

and for $c \in [a, b)$ the **right derivative** $f'_r(c)$ is defined by

$$f'_r(c) = \lim_{x \to c+} \frac{f(x) - f(c)}{x - c}.$$

a) Show that, if f is differentiable at c, it has both a left and a right derivative, and these are equal.

b) Determine the left and right derivatives at 0 of the function $x \mapsto |x(x - 1)|$.

4.3 Show that f, given by

$$f(x) = \begin{cases} x^2 \sin \dfrac{1}{x} & \text{if } x \neq 0 \\ 0 & \text{if } x = 0, \end{cases}$$

is differentiable at 0, but that f' is not continuous at 0.

4.4 In Exercise 3.21 you showed that if f and g are continuous at a point c then so are max $\{f, g\}$ and min $\{f, g\}$. Does this hold if we replace "continuous" by "differentiable"?

4.5 Let $f(x) = x^n$, where n is a **negative** integer. Show that $f'(x) = nx^{n-1}$.

4.2 The Mean Value Theorems

The following result, due to Rolle,[1] has far-reaching consequences:

Theorem 4.6 (Rolle's Theorem)

Let f be a real function, continuous on $[a, b]$ and differentiable on (a, b), and suppose that $f(a) = f(b)$. Then there exists c in (a, b) such that $f'(c) = 0$.

Proof

The result is visually obvious,

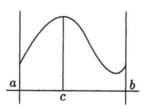

but by now we have seen enough strange functions to be suspicious of purely visual arguments. First, if $f(x) = f(a)$ for all x in $[a, b]$, then f is a constant function, and $f'(x) = 0$ for *every* c in (a, b). So suppose that there exists d in (a, b) such that $f(d) \neq f(a)$. Let us suppose in fact that $f(d) > f(a)$, for it

[1] Michel Rolle, 1652–1719

is easy to modify our argument to cope with the other case. By Theorem 3.19 there exists c in $[a, b]$ such that $f(c) = \sup_{[a,b]} f$. Since

$$f(c) \geq f(d) > f(a) = f(b)$$

it follows that $c \in (a, b)$. Then, for all $x < c$,

$$\frac{f(x) - f(c)}{x - c} \geq 0,$$

and so $f_l'(c) \geq 0$. Similarly, for all $x > c$,

$$\frac{f(x) - f(c)}{x - c} \leq 0,$$

and so $f_r'(c) \leq 0$. Since f is differentiable at c, we have (from Exercise 4.2) that $f_l'(c) = f_r'(c) = f'(c)$. Thus $f'(c) = 0$. $\square$

An immediate consequence of Rolle's Theorem is the following important result:

Theorem 4.7 (The (First) Mean Value Theorem)

Let f be a real function, continuous on $[a, b]$ and differentiable on (a, b). Then there exists c in (a, b) such that

$$f'(c) = \frac{f(b) - f(a)}{b - a}.$$

Proof

We apply Rolle's Theorem to the function $g(x) = f(x) - \lambda x$, where λ is a constant chosen so that $g(a) = g(b)$. Thus

$$f(a) - \lambda a = f(b) - \lambda b,$$

and so

$$\lambda = \frac{f(b) - f(a)}{b - a}.$$

Choosing c so that $g'(c) = f'(c) - \lambda = 0$, we see that

$$f'(c) = \lambda = \frac{f(b) - f(a)}{b - a},$$

as required. $\square$

Alternative statements of the First Mean Value Theorem are frequently useful:

With the conditions as stated, there exists c in (a, b) such that

$$f(b) = f(a) + (b - a)f'(c). \tag{4.5}$$

Let f be continuous in $[x, x + h]$ and differentiable in $(x, x + h)$; then there exists θ in $(0, 1)$ such that

$$f(x + h) = f(x) + hf'(x + \theta h). \tag{4.6}$$

This last version holds good even if h is negative.

We shall sometimes want to state that a function is **differentiable on** $[a, b]$. By this we shall mean that $f'(c)$ exists for all c in (a, b) and that $f'_r(a)$, $f'_l(b)$ exist.

One immediate consequence of the Mean Value Theorem is the following seemingly obvious, but in fact very important result:

Theorem 4.8

Let f be continuous on $[a, b]$ and differentiable on (a, b), and suppose that $f'(x) = 0$ for all x in (a, b). Then there exists a constant k such that $f(x) = k$ for all x in $[a, b]$.

Proof

Let $k = f(a)$ and let $c \in (a, b]$. Then, by Theorem 4.7, there exists d in (a, c) such that $f(c) = f(a) + (c - a)f'(d)$. Hence $f(c) = f(a) = k$ as required. $\square$

We now establish a generalisation of the Mean Value Theorem, due to Cauchy: [2]

Theorem 4.9 (Cauchy's Mean Value Theorem)

Let f, g be real functions, continuous on $[a, b]$ and differentiable on (a, b), and suppose that $g'(x) \neq 0$ throughout (a, b). Then there exists c in (a, b) such that

$$\frac{f'(c)}{g'(c)} = \frac{f(b) - f(a)}{g(b) - g(a)}.$$

---

[2] Augustin-Louis Cauchy, 1789–1847

Proof

Notice that Theorem 4.7 is a special case of this result, obtained by taking $g(x) = x$.

First, we observe that $g(b) \neq g(a)$, for if we had $g(b) = g(a)$ it would follow by Rolle's Theorem (4.6) that $g'(x) = 0$ for some x in (a, b), in contradiction to our assumptions. Now apply Rolle's Theorem to the function $h(x) = f(x) - \lambda g(x)$, where λ is a constant chosen so that $h(a) = h(b)$. Thus

$$\lambda = \frac{f(b) - f(a)}{g(b) - g(a)} .$$

Then, noting that h satisfies the conditions of Rolle's Theorem, we choose c in (a, b) such that $h'(c) = f'(c) - \lambda g'(c) = 0$. Since $g'(c) \neq 0$ we deduce that

$$\frac{f'(c)}{g'(c)} = \lambda = \frac{f(b) - f(a)}{b - a} ,$$

as required.   □

One useful consequence of this is a rule due to L'Hôpital: [3]

Theorem 4.10 (L'Hôpital's Rule)

Let f and g be real functions, differentiable in some interval $(c - \delta, c + \delta)$, and such that $f(c) = g(c) = 0$. If $\lim_{x \to c} \big(f'(x)/g'(x) \big)$ exists, then so does $\lim_{x \to c} \big(f(x)/g(x) \big)$, and

$$\lim_{x \to c} \frac{f(x)}{g(x)} = \lim_{x \to c} \frac{f'(x)}{g'(x)} .$$

Proof

Suppose that the latter limit exists and equals l. Then $f'(x)/g'(x)$ is defined in some set $(c - \delta_1, c + \delta_1) \setminus \{c\}$, where $0 < \delta_1 < \delta$, and so in particular $g'(x)$ is non-zero in that set. For each $\epsilon > 0$ there exists $\delta_2 > 0$ such that $0 < \delta_2 < \delta_1$ with the property that

$$\left| \frac{f'(x)}{g'(x)} - l \right| < \epsilon$$

for all x in $(c - \delta_2, c + \delta_2)$. If $x \in (c - \delta_2, c + \delta_2)$, then, by Theorem 4.9,

$$\frac{f(x)}{g(x)} = \frac{f(x) - f(c)}{g(x) - g(c)} = \frac{f'(\xi)}{g'(\xi)} ,$$

---

[3] Guillaume François Antoine, Marquis de L'Hôpital, 1661–1704

where ξ, being between x and c, certainly lies in the interval $(c - \delta_2, c + \delta_2)$. Hence

$$\left| \frac{f(x)}{g(x)} - l \right| = \left| \frac{f'(\xi)}{g'(\xi)} - l \right| < \epsilon.$$

Thus

$$\lim_{x \to c} \frac{f(x)}{g(x)} = l = \lim_{x \to c} \frac{f'(x)}{g'(x)},$$

as required.   □

Remark 4.11

It is not hard to modify the argument above to deal with left or right limits.

Example 4.12

Determine $\lim_{x \to 0}[(x - \sin x)/x^3]$.

Solution

We use L'Hôpital's Rule twice and the formula (3.15):

$$\lim_{x \to 0} \frac{x - \sin x}{x^3} = \lim_{x \to 0} \frac{1 - \cos x}{3x^2} = \lim_{x \to 0} \frac{\sin x}{6x} = \frac{1}{6}.$$

□

Theorem 4.13

Let f be continuous on $[a, b]$ and differentiable on (a, b). If $f'(x) > 0$ for all x in (a, b), then f is strictly increasing on $[a, b]$. If $f'(x) < 0$ for all x in (a, b), then f is strictly decreasing on $[a, b]$.

Proof

Suppose that $f'(x) > 0$ for all x in (a, b). Let $c, d \in [a, b]$, with $c < d$. Then, by Theorem 4.7, there exists ξ in (c, d) such that

$$f(d) = f(c) + (d - c)f'(\xi).$$

Thus $f(c) < f(d)$ and so f is increasing. The other statement follows in the same way.   □

Remark 4.14

The converse of this theorem is not quite true. For example, on the interval $[-1, 1]$ the function $f : x \mapsto x^3$ is strictly increasing, but $f'(0) = 0$.

EXERCISES

4.6 Determine
$$\lim_{x \to 0} \frac{\tan x - x}{x^3}, \quad \lim_{x \to 0} \frac{\sin x - x \cos x}{x^3}.$$

4.7 Let f be a function satisfying a **Hölder condition**[4] **of order** α at the point a. That is, suppose that there exist $M > 0$ and $\delta > 0$ such that, for all x in $(a - \delta, a + \delta) \setminus \{a\}$,
$$|f(x) - f(a)| < M|x - a|^\alpha.$$

a) Show that if $\alpha > 0$ then f is continuous at a, and that if $\alpha > 1$ then f is differentiable at a.

b) Give an example of a function f where $\alpha = 1$ but $f'(a)$ does not exist.

4.8 Let f be continuous on $[a, b]$ and differentiable on (a, b), and suppose that f attains its maximum and its minimum at points c, d, respectively. where $c, d \in (a, b)$. Show that $f'(c) = f'(d) = 0$.

4.9 Suppose that f, with domain $[a, b]$, has the property that $|f(x) - f(y)| \le (x - y)^2$ for all x, y in dom f. Show that f is a constant function.

4.10 Suppose that f has the property that $|f'(x)| < 1$ for all x in $(0, 1]$. Show that the sequence $(f(1/n))$ has a limit.

4.3 Inverse Functions

Let f be a real function, continuous throughout the interval $[a, b]$, differentiable in (a, b), and such that $f'(x) > 0$ for all x in (a, b). It follows that f is increasing and so, by Theorem 3.20, there is a continuous inverse function $f^{-1} : [f(a), f(b)]$ such that $f^{-1}(f(x)) = x$ for all x in $[a, b]$ and $f(f^{-1}(y)) = y$ for all y in $[f(a), f(b)]$.

[4] Otto Ludwig Hölder, 1859–1937

Theorem 4.15

Let f be a real function, continuous throughout the interval $[a, b]$, differentiable in (a, b), and such that $f'(x) > 0$ for all x in (a, b). Then f^{-1} is differentiable in $(f(a), f(b))$, and, for all d in $(f(a), f(b))$,

$$(f^{-1})'(d) = \frac{1}{f'(f^{-1}(d))} \, .$$

Proof

Let $d \in (f(a), f(b))$, let $c = f^{-1}(d)$, and let $\epsilon > 0$ be given. Since $f'(c) \neq 0$, we know that there exists $\delta_1 > 0$ such that

$$\left| \frac{x - c}{f(x) - f(c)} - \frac{1}{f'(c)} \right| < \epsilon \tag{4.7}$$

for all $x \neq c$ in (a, b) such that $|x - c| < \delta_1$. Since f^{-1} is continuous, there exists $\delta_2 > 0$ such that $|f^{-1}(y) - f^{-1}(d)| < \delta_1$ for all $y \neq d$ in $(f(a), f(b))$ such that $|y - d| < \delta_2$. Hence we can substitute $f^{-1}(y)$ for x in (4.7) to obtain

$$\left| \frac{f^{-1}(y) - f^{-1}(d)}{f(f^{-1}(y)) - f(f^{-1}(d))} - \frac{1}{f'(f^{-1}(d))} \right| < \epsilon$$

whenever $|y - d| < \delta_2$. That is,

$$\left| \frac{f^{-1}(y) - f^{-1}(d)}{y - d} - \frac{1}{f'(f^{-1}(d))} \right| < \epsilon$$

and from this we deduce that f^{-1} is differentiable at d, and that $(f^{-1})'(d) = 1/f'(f^{-1}(d))$. $\qquad\square$

The same result, with obvious changes, holds for a function with the property that $f'(x) > 0$ for all x in $[a, b]$.

Remark 4.16

The argument above is necessary to establish the differentiability of f^{-1}. If we assume the differentiability we can derive the formula quickly from the chain rule: since $f \circ f^{-1} = i$, the identity function, it follows that $(f \circ f^{-1})' = i' = 1$; hence

$$f'(f^{-1}(x))(f^{-1})'(x) = 1 \, ,$$

and so

$$(f^{-1})'(x) = \frac{1}{f'(f^{-1}(x))} \, .$$

As an application of this result, let us consider the function $f : x \rightarrow x^n$ ($n \in \mathbb{Z}$) in any interval $[a, b]$ such that $0 < a < b$. From (4.4) and Exercise 4.6 we know that $f'(x) = nx^{n-1}$. The inverse function, defined on $[a^{1/n}, b^{1/n}]$, is $f^{-1} : y \rightarrow y^{1/n}$. Then f^{-1} is differentiable, and

$$(f^{-1})'(y) = \frac{1}{f'(f^{-1}(y))} = \frac{1}{n(f^{-1}(y))^{n-1}} = \frac{1}{n(y^{1/n})^{n-1}} = \frac{1}{n}y^{\frac{1}{n}-1}. \quad (4.8)$$

Now let $q = m/n$ be an arbitrary rational number, and consider the function $f : x \mapsto x^q$. We see that $f = h \circ g$, where $g(x) = x^{1/n}$ and $h(u) = u^m$. It follows that

$$f'(x) = m(x^{1/n})^{m-1} \cdot \frac{1}{n}x^{(1/n)-1} = \frac{m}{n}x^{\frac{m-1}{n}+\frac{1}{n}-1} = qx^{q-1}.$$

Thus the validity of the formula nx^{n-1} for the derivative of the function $x \mapsto x^n$, originally obtained in (4.4) for positive integers n, extends to all rational exponents.

A further application is to the function $\sin : (-\pi/2, \pi/2) \rightarrow (-1, 1)$. As we saw in Section 3.7, there is an increasing inverse function $\sin^{-1} : [-1, 1] \rightarrow [-\pi/2, \pi/2]$. By Theorem 4.15, for all y such that $\cos(\sin^{-1} y) \neq 0$,

$$(\sin^{-1})'(y) = \frac{1}{\cos(\sin^{-1} y)} \cdot$$

Now, from $\cos^2 x + \sin^2 x = 1$ we deduce that $\cos x = \pm\sqrt{1 - \sin^2 x}$, and for x in $(-\pi/2, \pi/2)$ we can select the positive sign, since we know that $\cos x$ takes positive values in that interval. Substituting $\sin^{-1} y$ for x, and observing that $\sin(\sin^{-1} y) = y$, we see that $\cos(\sin^{-1} y) = \sqrt{1 - y^2}$. We have established part of the following theorem:

Theorem 4.17

Let $f : [-1, 1] \rightarrow [-\pi/2, \pi/2]$, $g : [-1, 1] \rightarrow [0, \pi]$ and $h : \mathbb{R} \rightarrow (-\pi/2, \pi/2)$ be given, respectively, by

$$f(x) = \sin^{-1} x, \quad g(x) = \cos^{-1} x, \quad h(x) = \tan^{-1} x.$$

Then

$$f'(x) = \frac{1}{\sqrt{1 - x^2}} \quad (x \in (-1, 1)), \qquad g'(x) = -\frac{1}{\sqrt{1 - x^2}} \quad (x \in (-1, 1)),$$

$$h'(x) = \frac{1}{1 + x^2} \quad (x \in \mathbb{R}).$$

Proof

The investigation of the derivative of $\cos^{-1}$ is essentially the same as for $\sin^{-1}$, and is left as an exercise. As for $\tan^{-1}$, we begin by noting that a simple use of the quotient rule for differentiation, applied to $\tan x = \sin x / \cos x$, gives the derivative $1/\cos^2 x$. Now

$$\frac{1}{\cos^2 x} = \frac{\cos^2 x + \sin^2 x}{\cos^2 x} = 1 + \tan^2 x,$$

and so we have

$$h'(y) = \frac{1}{1 + \tan^2(\tan^{-1} y)} = \frac{1}{1 + y^2},$$

as required.   □

Remark 4.18

The derivative of sin is 0 at $\pm\pi/2$, and so $\sin^{-1}$, though defined at ± 1, does not have a derivative at those points. A similar remark applies to $\cos^{-1}$.

EXERCISES

4.11 Show that the derivative for $\cos^{-1} x$, for x in $(-1, 1)$, is $-1/\sqrt{1-x^2}$.

4.12 Give another proof of the result of Exercise 3.34, that $\cos^{-1} x + \sin^{-1} x = \pi/2$ for all x in $[-1, 1]$.

4.13 Let $f : \mathbb{R} \to (0, \infty)$ have the property that $f'(x) = f(x)$ for all x. Show that f is an increasing function for all x. Show also that $(f^{-1})'(x) = 1/x \quad (x > 0)$.

4.4 Higher Derivatives

In many cases it is possible to repeat the operation of differentiation by applying the process to the derivative. It is reasonable to write the function $(f')'$ more simply as f'', and to call it the **second derivative** of f. Similarly we have the **third derivative** f''', and we may repeat the process as long as the functions remain differentiable. The nth derivative is denoted by $f^{(n)}$, and sometimes, for consistency, we shall want to use that notation even for $n = 0, 1, 2, 3$. It is worth mentioning that other notations for $f^{(n)}(x)$ include

$$(D_x^n f)(x) \quad \text{and} \quad \frac{d^n}{dx^n} f(x).$$

Let f and g be real functions that are n times differentiable. From Theorem 4.2 we know that $(f + g)' = f' + g'$, and from this we easily deduce that

$$(f + g)'' = (f' + g')' = f'' + g''.$$

More generally, we have

$$(f + g)^{(n)} = f^{(n)} + g^{(n)} \quad (n \geq 1). \tag{4.9}$$

Similarly, if k is constant, then

$$(kf)^{(n)} = kf^{(n)}. \tag{4.10}$$

The product of f and g presents more of a problem. Recall from Theorem 4.2 that

$$(f \cdot g)' = f' \cdot g + f \cdot g',$$

from which we easily see that

$$\begin{aligned}(f \cdot g)'' &= (f' \cdot g + f \cdot g')' = (f' \cdot g)' + (f \cdot g')' \\ &= (f'' \cdot g + f' \cdot g') + (f' \cdot g' + f \cdot g'') = f'' \cdot g + 2f' \cdot g' + f \cdot g''.\end{aligned}$$

A similar calculation establishes that

$$(f \cdot g)''' = f''' \cdot g + 3f'' \cdot g' + 3f' \cdot g'' + f \cdot g''',$$

and these results suggest that we might be able to establish a formula akin to the binomial theorem. The following result is due to Leibniz. (In the formula we interpret $f^{(0)}$ and $g^{(0)}$ in the obvious way as f and g.)

Theorem 4.19 (Leibniz's Theorem)

Let $n \geq 1$, and let f and g be real functions, each differentiable n times. Then

$$(f \cdot g)^{(n)} = \sum_{r=0}^{n} \binom{n}{r} f^{(n-r)} \cdot g^{(r)}.$$

Proof

We know that the result holds for $n = 1$ and $n = 2$. Let $n \geq 2$, and suppose inductively that the result holds for $n - 1$. Then

$$(f \cdot g)^{(n)} = \left((f \cdot g)^{(n-1)}\right)' = \left(\sum_{r=0}^{n-1} \binom{n-1}{r} f^{(n-1-r)} \cdot g^{(r)}\right)'.$$

For $r = 1, 2, \ldots, n - 1$ we have terms

$$\binom{n-1}{r-1} f^{(n-r)} \cdot (g^{(r-1)})' + \binom{n-1}{r} (f^{(n-1-r)})' \cdot g^{(r)} \,.$$

The sum of these terms is

$$\left[\binom{n-1}{r-1} + \binom{n-1}{r} \right] f^{(n-r)} \cdot g^{(r)} \,,$$

and this is equal to

$$\binom{n}{r} f^{(n-r)} \cdot g^{(r)}$$

by the Pascal Triangle Identity (1.14). In addition we have terms

$$\binom{n-1}{0} f^{(n)} \cdot g^{(0)} = \binom{n}{0} f^{(n)} \cdot g^{(0)}, \quad \binom{n-1}{n-1} f^{(0)} \cdot g^{(n)} = \binom{n}{n} f^{(0)} \cdot g^{(n)} \,,$$

and so

$$(f \cdot g)^{(n)} = \sum_{r=0}^{n} \binom{n}{r} f^{(n-r)} \cdot g^{(r)} \,,$$

as required.　　　　　　　　　　　　　　　　　　　　□

This is of course a rather cumbersome formula, and is of use mainly when one of the functions has the property that $f^{(r)} = 0$ for sufficiently large r.

Example 4.20

Find the $2n$th derivative of the function $h : \mapsto x^2 \sin x$.

Solution

By Leibniz's Theorem,

$$h^{(2n)}(x) = x^2 \sin^{(2n)}(x) + 2n.2x. \sin^{(2n-1)}(x) + \binom{2n}{2} .2. \sin^{(2n-2)}(x) \,,$$

since all subsequent terms vanish. Now, for all $k \geq 0$,

$$\sin^{(2k)}(x) = (-1)^k \sin x \,, \quad \sin^{2k-1}(x) = (-1)^{k-1} \cos x \,.$$

Hence, collecting terms, we obtain

$$h^{2n}(x) = (-1)^{n-1}[(4n^2 - 2n - x^2) \sin x + 4nx \cos x] \,.$$

　　　　　　　　　　　　　　　　　　　　□

EXERCISES

4.14 Show that, for a twice differentiable function f,

$$\lim_{h \to 0} \frac{f(a+2h) - 2f(a+h) + f(a)}{h^2} = f''(a).$$

4.15 Determine a formula for the $(2n-1)$th derivative of $x^2 \cos x$.

4.16 Let $f(x) = \cos(m \sin^{-1} x)$. Show by induction on n that, for all $n \geq 0$,

$$(1-x^2)f^{(n+2)}(x) - (2n+1)xf^{(n+1)}(x) + (m^2 - n^2)f^{(n)}(x) = 0.$$

Deduce that, for all even n,

$$f^{(n)}(0) = [(n-2)^2 - m^2]\ldots[2^2 - m^2][-m^2].$$

4.5 Taylor's Theorem

In this section we establish a generalization of the First Mean Value Theorem (Theorem 4.7) due to Taylor. [5]

Theorem 4.21 (Taylor's Theorem)

Let $n \in \mathbb{N}$. Suppose that f is n times differentiable in $[a,b]$. Then there exists c in (a,b) such that

$$f(b) = f(a) + (b-a)f'(a) + \frac{(b-a)^2}{2!}f''(a) + \cdots + \frac{(b-a)^{n-1}}{(n-1)!}f^{(n-1)}(a)$$
$$+ \frac{(b-a)^n}{n!}f^{(n)}(c).$$

Proof

By way of preparation, let us consider the function F_n defined on $[a,b]$ by

$$F_n(x) = f(b) - f(x) - (b-x)f'(x) - \frac{(b-x)^2}{2!}f''(x) - \cdots - \frac{(b-x)^{n-1}}{(n-1)!}f^{(n-1)}(x).$$

[5] Brook Taylor, 1685–1731

Then

$$F_n'(x) = -f'(x) + [f'(x) - (b-x)f''(x)]$$
$$+ \left[(b-x)f''(x) - \frac{1}{2!}(b-x)^2 f'''(x)\right]$$
$$+ \cdots + \left[\frac{(b-x)^{n-2}}{(n-2)!}f^{(n-1)}(x) - \frac{(b-x)^{n-1}}{(n-1)!}f^{(n)}(x)\right]$$
$$= -\frac{(b-x)^{n-1}}{(n-1)!}f^{(n)}(x), \tag{4.11}$$

since all other terms cancel.

Now consider the function G_n defined on $[a,b]$ by

$$G_n(x) = F_n(x) - \left(\frac{b-x}{b-a}\right)^n F_n(a),$$

and observe that $G_n(a) = G_n(b) = 0$. Since G_n is continuous on $[a,b]$ and differentiable on (a,b), we can apply Rolle's Theorem (Theorem 4.6) and assert that there exists c in (a,b) such that $G_n'(c) = 0$. That is, by (4.11),

$$0 = G_n'(c) = -\frac{(b-c)^{n-1}}{(n-1)!}f^{(n)}(c) + \frac{n(b-c)^{n-1}}{(b-a)^n}F_n(a)$$
$$= \frac{n(b-c)^{n-1}}{(b-a)^n}\left[F_n(a) - \frac{(b-a)^n}{n!}f^{(n)}(c)\right]$$
$$= \frac{n(b-c)^{n-1}}{(b-a)^n}\left[f(b) - f(a) - (b-a)f'(a) - \cdots\right.$$
$$\left. - \frac{(b-a)^{n-1}}{(n-1)!}f^{(n-1)}(a) - \frac{(b-a)^n}{n!}f^{(n)}(c)\right],$$

and it immediately follows that

$$f(b) = f(a) + (b-a)f'(a) + \cdots + \frac{(b-a)^{n-1}}{(n-1)!}f^{(n-1)}(a) + \frac{(b-a)^n}{n!}f^{(n)}(c),$$

exactly as required. $\qquad\square$

A small change of notation, in which b is written as $a+h$, gives

$$f(a+h) = f(a) + hf'(a) + \cdots + \frac{h^{n-1}}{(n-1)!}f^{(n-1)}(a) + \frac{h^n}{n!}f^{(n)}(a+\theta h) \tag{4.12}$$

where $0 < \theta < 1$, and yet another version, often called **Maclaurin's**[6] **Theorem**, is obtained if we put $a = 0$ and replace h by x:

$$f(x) = f(0) + xf'(0) + \cdots + \frac{x^{n-1}}{(n-1)!}f^{(n-1)}(0) + R_n, \tag{4.13}$$

---

[6] Colin Maclaurin, 1698–1746

where R_n, the **remainder term**, has the value

$$\frac{x^n}{n!} f^{(n)}(\theta x),$$

for some θ in $(0, 1)$.

If, in (4.13), $R_n \to 0$ as $n \to \infty$ for suitable values of x, we obtain the **Taylor–Maclaurin series**

$$f(x) = \sum_{n=0}^{\infty} \frac{x^n}{n!} f^{(n)}(0),$$

and we say that the function is **analytic**. Such a function is in particular **infinitely differentiable**, but we shall see in Exercise 6.17 that it is possible for a function to be infinitely differentiable but not analytic.

We shall return to the consideration of Taylor–Maclaurin series in a more general context in Chapter 7.

EXERCISES

4.17 Show that the Taylor–Maclaurin series of a polynomial

$$a_0 + a_1 x + \cdots + a_n x^n$$

is precisely that polynomial.

5.1 The Riemann Integral

It is possible from a traditional calculus course to gain the impression that integration is simply "anti-differentiation". This, as we shall see, is part of the message, but it is not the main issue. Integration arose as a limiting case of a sum, where the individual summands tend to zero and the number of summands tends to infinity, and it is this aspect that is fundamental.

The first really satisfactory theory of integration was developed by Riemann, and it is his theory that is presented here. Let f be a bounded real function on a closed interval $[a, b]$. A **dissection** D of $[a, b]$ is a set $\{x_0, x_1, \ldots, x_n\}$ with the properties that $n \geq 1$, $x_0 = a$, $x_n = b$, and $x_{i-1} < x_i$ $(i = 1, 2, \ldots, n)$. Let $\mathcal{D}[a, b]$ denote the set of all dissections of $[a, b]$. Since f is bounded both above and below in each of the open intervals (x_{i-1}, x_i), we may define, for $i = 1, 2, \ldots, n$,

$$\sup_{(x_{i-1}, x_i)} f = M_i, \qquad \inf_{(x_{i-1}, x_i)} f = m_i.$$

The **upper sum of f relative to the dissection D** is defined by

$$\mathcal{U}(f, D) = \sum_{i=1}^{n} M_i(x_i - x_{i-1}).$$

The **lower sum of f relative to the dissection D** is defined in a similar way by

$$\mathcal{L}(f, D) = \sum_{i=1}^{n} m_i(x_i - x_{i-1}).$$

It is clear that

$$\mathcal{L}(f, D) \leq \mathcal{U}(f, D). \tag{5.1}$$

Now let $M = \sup_{[a,b]} f$, $m = \inf_{[a,b]} f$. Then $M_i \leq M$ for all i, and so

$$\mathcal{U}(f, D) \leq \sum_{i=1}^{n} M(x_i - x_{i-1}) = M \sum_{i=1}^{n}(x_i - x_{i-1}) = M(b - a). \tag{5.2}$$

Similarly,

$$\mathcal{L}(f, D) \geq m(b - a). \tag{5.3}$$

It follows from (5.1) and (5.3) that the set $\{\mathcal{U}(f, D) : D \in \mathcal{D}[a, b]\}$ is bounded below by $m(b - a)$, and so we may define the **upper integral of f over** $[a, b]$ by

$$\overline{\int_a^b} f = \inf \{\mathcal{U}(f, D) : D \in \mathcal{D}[a, b]\}. \tag{5.4}$$

In a similar way we define the **lower integral of f over** $[a, b]$ by

$$\underline{\int_a^b} f = \sup \{\mathcal{L}(f, D) : D \in \mathcal{D}[a, b]\}. \tag{5.5}$$

The diagram

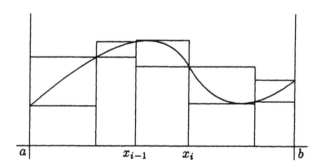

demonstrates the intuition behind these formal definitions. The lower sum approximates to the area under the curve by means of the lower rectangles, and the upper sum by means of the upper rectangles.

For "reasonable" functions one might expect that the lower and upper sums will approach each other ever more closely as the size of the subintervals $[x_{i-1}, x_i]$ decreases, and that the difference between the two sums will tend to zero. Our task now is to make this intuition mathematically precise. We can begin with a definition, that f is **Riemann integrable over** $[a, b]$ if

$$\overline{\int_a^b} f = \underline{\int_a^b} f \, ;$$

in this case we denote the common value of the upper and lower integrals by

$$\int_a^b f, \quad \text{or by} \quad \int_a^b f(x) \, dx \, .$$

Intuitively, we think of $\int_a^b f$ as the (signed) area between the graph of f and the x-axis between the lines $x = a$ and $x = b$ – where "signed" means that areas below the x-axis are to be regarded as negative.

An alternative notation for $\int_a^b f$ is $\int_a^b f(x) \, dx$. This notation is useful if our function f is given by an explicit expression, such as $x^2 - x + 2$, when it is very convenient to write $\int_a^b (x^2 - x + 2) \, dx$. It is important to realise, however, that x is here a "dummy", in the sense that the expression $\int_a^b f(x) \, dx$ does not depend on x: the expression $\int_a^b f(u) \, du$ has exactly the same meaning.

The first important result in the theory is given in the following theorem:

Theorem 5.1

Let f be a bounded function on the interval $[a, b]$. Then

$$\underline{\int_a^b} f \le \overline{\int_a^b} f \, .$$

Proof

We have already observed the obvious result that

$$\mathcal{L}(f, D) \le \mathcal{U}(f, D)$$

for every dissection D. We now establish the less obvious result that *every* lower sum is less than or equal to *every* upper sum. To be more precise, we show that for arbitrarily chosen dissections D_1, D_2 in $\mathcal{D}[a, b]$,

$$\mathcal{L}(f, D_1) \le \mathcal{U}(f, D_2).$$

Let us say that a dissection D' is a **refinement** of D if $D \subseteq D'$. Then we have the following lemma, which deals with the simplest sort of refinement, obtained by adding just one point:

Lemma 5.2

Let $D = \{a = x_0, x_1, \ldots, x_n = b\}$ be a dissection, and let $D' = D \cup \{y\}$, where $y \notin D$. Then

$$\mathcal{U}(f, D') \leq \mathcal{U}(f, D), \quad \mathcal{L}(f, D') \geq L(f, D).$$

Proof

Suppose that $x_{i-1} < y < x_i$, and let

$$P_i = \sup_{(x_{i-1}, y)} f, \; p_i = \inf_{(x_{i-1}, y)} f, \quad Q_i = \sup_{(y, x_i)} f, \; q_i = \inf_{(y, x_i)} f.$$

Then, if as usual we write

$$M_i = \sup_{(x_{i-1}, x_i)} f, \quad m_i = \inf_{(x_{i-1}, x_i)} f,$$

we easily see that $P_i, Q_i \leq M_i$, $p_i, q_i \geq m_i$. Since $\mathcal{U}(f, D)$ and $\mathcal{U}(f, D')$ differ only in the contribution from the subinterval $[x_{i-}, x_i]$,

$$\mathcal{U}(f, D') - \mathcal{U}(f, D) = P_i(y - x_{i-1}) + Q_i(x_i - y) - M_i(x_i - x_{i-1})$$
$$\leq M_i(y - x_{i-1}) + M_i(x_i - y) - M_i(x_i - x_{i-1}) = 0.$$

A similar argument shows that $\mathcal{L}(f, D') - \mathcal{L}(f, D) \geq 0$.   □

We use this lemma to prove a more general result:

Lemma 5.3

If D' is a refinement of D, then

$$\mathcal{U}(f, D') \leq \mathcal{U}(f, D), \quad \mathcal{L}(f, D') \geq L(f, D). \tag{5.6}$$

Proof

The proof, by induction on $m = |D'| - |D|$, begins with the observation that, by virtue of Lemma 5.2, the inequalities (5.6) are valid for $m = 1$. Suppose that the inequalities hold for every dissection D and for every refinement D' of D such that $|D'| - |D| = m - 1$. If we now suppose that D' is a refinement of D with $|D'| - |D| = m$, we can choose y in $D' \setminus D$, and define D'' as $D' \setminus \{y\}$. By our induction hypothesis,

$$\mathcal{U}(f, D'') \leq \mathcal{U}(f, D), \quad \mathcal{L}(f, D'') \geq L(f, D).$$

Then, since $D' = D'' \cup \{y\}$, it follows from Lemma 5.2 that

$$\mathcal{U}(f, D') \leq \mathcal{U}(f, D'') \leq \mathcal{U}(f, D), \quad \mathcal{L}(f, D') \geq \mathcal{L}(f, D'') \geq \mathcal{L}(f, D).$$

$\square$

The theorem now follows easily, since $D_1 \cup D_2$ is a common refinement of D_1 and D_2, and so

$$\mathcal{L}(f, D_1) \leq \mathcal{L}(f, D_1 \cup D_2) \leq \mathcal{U}(f, D_1 \cup D_2) \leq \mathcal{U}(f, D_2).$$

We may therefore conclude that

$$\underline{\int_a^b} f \leq \mathcal{U}(f, D)$$

for *every* dissection D in $\mathcal{D}[a, b]$, and so

$$\underline{\int_a^b} f \leq \inf \{\mathcal{U}(f, D) : D \in \mathcal{D}[a, b]\} = \overline{\int_a^b} f.$$

$\square$

Before proceeding further, we look at three examples.

Example 5.4

Show that the function $f : x \mapsto x$ is Riemann integrable in $[0, 1]$, and that $\int_0^1 f = \frac{1}{2}$.

Solution

Let D_n be the dissection $\{0, \frac{1}{n}, \frac{2}{n}, \ldots, \frac{n-1}{n}, 1\}$. In the subinterval $[\frac{i-1}{n}, \frac{i}{n}]$ $(i = 1, 2, \ldots, n)$ the supremum and infimum of f are given by

$$M_i = \frac{i}{n}, \quad m_i = \frac{i-1}{n}.$$

Hence

$$\mathcal{U}(f, D_n) = \sum_{i=1}^{n} \frac{i}{n} \cdot \frac{1}{n} = \frac{1}{n^2}(1 + 2 + \ldots + n)$$

$$= \frac{n(n+1)}{2n^2} = \frac{1}{2} + \frac{1}{2n},$$

and

$$\mathcal{L}(f, D_n) = \sum_{i=1}^{n} \frac{i-1}{n} \cdot \frac{1}{n} = \frac{1}{n^2} \left(0 + 1 + \ldots + (n-1) \right)$$

$$= \frac{n(n-1)}{2n^2} = \frac{1}{2} - \frac{1}{2n}.$$

Now

$$\overline{\int_a^b} f \le \mathcal{U}(f, D_n), \quad \underline{\int_a^b} f \ge \mathcal{L}(f, D_n),$$

and so, using Theorem 5.1, we have

$$0 \le \overline{\int_a^b} f - \underline{\int_a^b} f \le \mathcal{U}(f, D_n) - \mathcal{L}(f, D_n) = \frac{1}{n}.$$

Since this holds for *every* n, we deduce that

$$\overline{\int_a^b} f = \underline{\int_a^b} f,$$

and so f is Riemann integrable. Since we also know that

$$\frac{1}{2} - \frac{1}{2n} \le \int_a^b f \le \frac{1}{2} + \frac{1}{2n}$$

for every $n \ge 1$, we may conclude that $\int_a^b f = \frac{1}{2}$.

Example 5.5

Let f be the function on $[a, b]$ given by

$$f(x) = \begin{cases} 0 & \text{if } x \text{ is rational} \\ 1 & \text{if } x \text{ is irrational.} \end{cases}$$

Show that f is not Riemann integrable.

Solution

For every dissection $D = \{a, x_1, \ldots, x_{n-1}, b\}$, every subinterval (x_{i-1}, x_i) contains both rational and irrational points, and so, for $i = 1, 2, \ldots, n$,

$$M_i = 1, \quad m_i = 0.$$

Hence, for every dissection D,

$$\mathcal{U}(f, D) = b - a, \quad \mathcal{L}(f, D) = 0,$$

and we deduce that

$$\overline{\int_a^b} f = b - a, \qquad \underline{\int_a^b} f = 0.$$

Thus f is not Riemann integrable.   □

It is perhaps not too surprising that such a bizarre function fails to be Riemann integrable. The next example, which revisits the function first encountered in Example 3.9, shows, however, that some very strange functions can be Riemann integrable:

Example 5.6

Let f, with domain $[0, 1]$, be defined by

$$f(x) = \begin{cases} 0 & \text{if } x \text{ is irrational} \\ 1/q & \text{if } x = p/q \text{ is a rational expressed in lowest terms.} \end{cases}$$

Show that f is Riemann integrable on $[0, 1]$, and that $\int_0^1 f = 0$.

Solution

For each $n \geq 2$ let D_n be the dissection

$$\{p/q \in [0, 1] \,:\, p \leq q \leq n\}.$$

Thus, for example, $D_4 = \{0, \frac{1}{4}, \frac{1}{3}, \frac{1}{2}, \frac{2}{3}, \frac{3}{4}, 1\}$.

For each interval $[x_{i-1}, x_i]$, we have

$$m_i = \inf_{(x_{i-1}, x_i)} f = 0, \qquad \sup_{(x_{i-1}, x_i)} f < \frac{1}{n},$$

and from this it follows that

$$\mathcal{U}(f, D_n) - \mathcal{L}(f, D_n) < \frac{1}{n} \sum_i (x_i - x_{i-1}) = \frac{1}{n}.$$

We conclude that

$$\overline{\int_0^1} f - \underline{\int_0^1} f \leq \frac{1}{n},$$

and, since this happens for all n we must have

$$\overline{\int_0^1} f = \underline{\int_0^1} f.$$

Thus f is Riemann integrable on $[0, 1]$.

Finally, it is clear that $\mathcal{L}(f, D) = 0$ for every dissection D. Hence

$$\int_0^1 f = \underline{\int_0^1} f = 0.$$

□

EXERCISES

5.1 Show that the constant function $C_k : x \mapsto k$ is Riemann integrable in every interval $[a, b]$, and that $\int_a^b C_k = k(b - a)$.

5.2 Show, using the technique of Example 5.4 , that $f : x \mapsto x^2$ is Riemann integrable in $[0, 1]$, and that $\int_0^1 x^2 \, dx = 1/3$.

5.3 Let $c \in (0, 1)$, and let f, with domain $[0, 1]$, be given by

$$f(x) = \begin{cases} 1 & \text{if } x = c \\ 0 & \text{otherwise.} \end{cases}$$

Show that f is Riemann integrable, and that $\int_0^1 f = 0$.

5.2 Classes of Integrable Functions

In this section we shall see that some quite broad classes of functions are Riemann integrable, and it will be convenient to write "$f \in \mathcal{R}[a, b]$" as a shorthand for "f is Riemann integrable in $[a, b]$". The following lemma is useful:

Lemma 5.7

Let f, with domain $[a, b]$, be a bounded function. Then $f \in \mathcal{R}[a, b]$ if and only if, for every $\epsilon > 0$ there is a dissection D in $\mathcal{D}[a, b]$ such that

$$0 \leq \mathcal{U}(f, D) - \mathcal{L}(f, D) < \epsilon.$$

Proof

Suppose first that $f \in \mathcal{R}[a, b]$. Let $\epsilon > 0$ be given. Since

$$\int_a^b f = \inf \{\mathcal{U}(f, D) : D \in \mathcal{D}[a, b]\} = \sup \{\mathcal{L}(f, D) : D \in \mathcal{D}[a, b]\},$$

there exist dissections D_1, D_2 such that

$$0 \leq \mathcal{U}(f, D_1) - \int_a^b f < \frac{\epsilon}{2}, \quad 0 \leq \int_a^b f - \mathcal{L}(f, D_2) < \frac{\epsilon}{2}.$$

Let $D = D_1 \cup D_2$, a common refinement of D_1 and D_2. Then, by Lemma 5.3, $\mathcal{U}(f, D) \leq \mathcal{U}(f, D_1)$ and $\mathcal{L}(f, D) \geq \mathcal{L}(f, D_2)$, and so

$$0 \leq \mathcal{U}(f, D) - \mathcal{L}(f, D) \leq \mathcal{U}(f, D_1) - \mathcal{L}(f, D_2)$$

$$\leq \left[\mathcal{U}(f, D_1) - \int_a^b f\right] + \left[\int_a^b f - \mathcal{L}(f, D_2)\right]$$

$$< \epsilon.$$

Conversely, let $\epsilon > 0$, and suppose that there is a dissection D_ϵ in $\mathcal{D}[a, b]$ such that

$$0 \leq \mathcal{U}(f, D_\epsilon) - \mathcal{L}(f, D_\epsilon) < \epsilon.$$

Then

$$\overline{\int_a^b} f \leq \mathcal{U}(f, D_\epsilon) < \mathcal{L}(f, D_\epsilon) + \epsilon \leq \underline{\int_a^b} f + \epsilon,$$

and so

$$\overline{\int_a^b} f - \underline{\int_a^b} f < \epsilon.$$

This holds for *every* positive ϵ, and so we must have

$$\overline{\int_a^b} f = \underline{\int_a^b} f.$$

That is, $f \in \mathcal{R}[a, b]$. □

Recall that a function f defined on $[a, b]$ is **(monotonic) increasing** if, for all x, y in $[a, b]$,

$$x \leq y \implies f(x) \leq f(y)$$

and **(monotonic) decreasing** if, again for all x, y in $[a, b]$,

$$x \leq y \implies f(x) \geq f(y).$$

The term **monotonic** is used for a function that is either increasing or decreasing.

Theorem 5.8

Let f be a monotonic function on $[a, b]$. Then f is Riemann integrable on $[a, b]$.

Proof

It will be sufficient to consider the case where f is monotonic increasing. We remark that $f(a) \leq f(x) \leq f(b)$ for all x in $[a, b]$, and so certainly f is bounded. Consider the dissection $D_n = \{x_0, x_1, \ldots, x_n\}$ that partitions $[a, b]$ into n equal subintervals:

$$x_i = a + \frac{i}{n}(b - a) \quad (i = 0, 1, \ldots, n).$$

The monotonic property implies that, in our usual notation,

$$M_i \le f(x_i), \quad m_i \ge f(x_{i-1}),$$

and so, since every subinterval has fixed length $(b-a)/n$,

$$\mathcal{U}(f, D_n) \le \frac{b-a}{n} \big(f(x_1) + f(x_2) + \cdots + f(x_{n-1}) + f(b) \big),$$

$$\mathcal{L}(f, D_n) \ge \frac{b-a}{n} \big(f(a) + f(x_1) + \cdots + f(x_{n-2}) + f(x_{n-1}) \big).$$

Hence

$$\mathcal{U}(f, D_n) - \mathcal{L}(f, D_n) \le \frac{(b-a)(f(b) - f(a))}{n}.$$

For a given $\epsilon > 0$ we now choose $n > [(b-a)(f(b) - f(a))]/\epsilon$, thus obtaining

$$\mathcal{U}(f, D_n) - \mathcal{L}(f, D_n) < \epsilon.$$

From Lemma 5.7 we deduce that f is Riemann integrable on $[a, b]$. $\square$

Theorem 5.9

If f is continuous on $[a, b]$, then it is Riemann integrable on $[a, b]$.

Proof

Suppose that f is continuous on $[a, b]$. By Theorem 3.17 it is uniformly continuous, and so for each $\epsilon > 0$ there exists $\delta > 0$ such that, for all x, y in $[a, b]$,

$$|x - y| < \delta \implies |f(x) - f(y)| < \frac{\epsilon}{3(b-a)}.$$

Consider now the dissection $D = \{x_0, x_1, \ldots, x_n\}$ that divides $[a, b]$ into n equal subintervals, each of length $(b-a)/n$, and choose n so that $(b-a)/n < \delta$. For each i in $\{1, 2, \ldots, n\}$ we can choose points x, y in (x_{i-1}, x_i) such that

$$f(x) > M_i - \frac{\epsilon}{3(b-a)}, \quad f(y) < m_i + \frac{\epsilon}{3(b-a)}.$$

Certainly $|x - y| < (b-a)/n < \delta$, and so

$$|f(x) - f(y)| < \frac{\epsilon}{3(b-a)}.$$

Hence

$$\begin{aligned}
M_i - m_i &= \big(M_i - f(x) \big) + \big(f(x) - f(y) \big) + \big(f(y) - m_i \big) \\
&\le \big(M_i - f(x) \big) + |f(x) - f(y)| + \big(f(y) - m_i \big) \\
&< \frac{\epsilon}{3(b-a)} + \frac{\epsilon}{3(b-a)} + \frac{\epsilon}{3(b-a)} = \frac{\epsilon}{b-a}.
\end{aligned}$$

Hence

$$\mathcal{U}(f, D) - \mathcal{L}(f, D) = \sum_{i=1}^{n} (M_i - m_i)(x_i - x_{i-1})$$

$$< \frac{\epsilon}{b-a} \sum_{i=1}^{n} (x_i - x_{i-1}) = \epsilon,$$

and the result now follows from Lemma 5.7.   □

We have already seen in Example 5.6 that the class of Riemann integrable functions includes some fairly strange items not covered by Theorems 5.8 and 5.9. We finish this section with another example concerning a function we have encountered before:

Example 5.10

Let f be the function defined on $[-1, 1]$ by

$$f(x) = \begin{cases} \sin(1/x) & \text{if } x \neq 0 \\ 0 & \text{if } x = 0. \end{cases}$$

Show that $f \in \mathcal{R}[-1, 1]$.

Solution

We know that f is not continuous at 0. Since it is certainly not monotonic, neither Theorem 5.8 nor Theorem 5.9 applies. We make use of Lemma 5.7. Let $\epsilon > 0$. We form a dissection

$$D = \{x_0 = -1, x_1, \ldots, x_m, -\frac{\epsilon}{8}, \frac{\epsilon}{8}, x_{m+1}, \ldots, x_n = 1\}.$$

Since f is continuous, and so integrable, in the interval $[-1, -\epsilon/8]$, we can arrange for the dissection

$$D_1 = \{x_0, \ldots, x_m, -\frac{\epsilon}{8}\}$$

to be such that

$$\mathcal{U}(f, D_1) - \mathcal{L}(f, D_1) < \frac{\epsilon}{4}.$$

Similarly, we can arrange for the dissection

$$D_2 = \{\frac{\epsilon}{8}, x_{m+1}, \ldots, x_n\}$$

to be such that

$$\mathcal{U}(f, D_2) - \mathcal{L}(f, D_2) < \frac{\epsilon}{4}.$$

Now, within the interval $(-(\epsilon/8), \epsilon/8)$ the function has supremum 1 and infimum -1. So the contribution of this subinterval to $\mathcal{U}(f, D) - \mathcal{L}(f, D)$ is $(1 - (-1)) \times (\epsilon/4) = \epsilon/2$. Hence

$$\mathcal{U}(f, D) - \mathcal{L}(f, D)$$
$$= \mathcal{U}(f, D_1) - \mathcal{L}(f, D_1) + \frac{\epsilon}{2} + \mathcal{U}(f, D_2) - \mathcal{L}(f, D_2)$$
$$< \frac{\epsilon}{4} + \frac{\epsilon}{2} + \frac{\epsilon}{4} = \epsilon,$$

and so $f \in \mathcal{R}[-1, 1]$.   □

Remark 5.11

This technique, of "pinching out" an isolated discontinuity, is generally applicable. See Exercise 5.4.

EXERCISES

5.4 Let f, with domain $[a, b]$, be continuous except at finitely many points $c_1, c_2, \ldots, c_k$. Show that $f \in \mathcal{R}[a, b]$.

5.5 Let f be continuous on $[a, b]$, and let $f(x) \geq 0$ for all x in $[a, b]$. Show that, if there exists c in $[a, b]$ such that $f(c) > 0$, then $\int_a^b f > 0$.

5.6 Let f be continuous on $[a, b]$, and suppose that $\int_a^b (f(x))^2 \, dx = 0$. Show that $f(x) = 0$ for all x in $[a, b]$.

5.7 Show that if $f \in \mathcal{R}[a, b]$ then $f \in \mathcal{R}[c, d]$ for all c, d such that $a \leq c < d \leq b$.

5.8 Let $D_n = \{x_0, x_1, \ldots, x_n\}$ denote the dissection of $[a, b]$ in which the interval is divided into n equal parts:

$$x_i = a + \frac{i}{n}(b - a).$$

Show that, if f is continuous on $[a, b]$, $\mathcal{U}(f, D_n) - \mathcal{L}(f, D_n)$ can be made less than any given ϵ by taking n sufficiently large. Deduce that the sequences $(\mathcal{U}(f, D_n))$ and $(\mathcal{L}(f, D_n))$ both have limit $\int_a^b f$.

5.9 Consider the function f defined on $[0, 1]$ by the rule that $f(0) = 0$ and, for all $m \geq 0$,

$$f(x) = \frac{(-1)^m}{2^m} \quad \text{when} \quad \frac{1}{2^{m+1}} < x \leq \frac{1}{2^m}.$$

Show that $f \in \mathcal{R}[a, b]$, and find the value of $\int_a^b f$.

5.3 Properties of Integrals

This rather technical section establishes some of the fundamental properties of integration. Most are intuitively "obvious" if we regard an integral as an area under a curve.

Theorem 5.12

Let $a, b, c \in \mathbb{R}$, and suppose that $a < b < c$. Let f be a bounded function whose domain contains $[a, c]$. Then $f \in \mathcal{R}[a, c]$ if and only if $f \in \mathcal{R}[a, b]$ and $f \in \mathcal{R}[b, c]$. If the integrals exist, then

$$\int_a^c f = \int_a^b f + \int_b^c f. \tag{5.7}$$

Proof

Let $\epsilon > 0$ be given. Suppose first that $f \in \mathcal{R}[a, c]$, and let D be a dissection of $[a, c]$ such that
$$\mathcal{U}(f, D) - \mathcal{L}(f, D) < \epsilon.$$

The dissection D may or may not contain b. The dissection $D' = D \cup \{b\}$ is either equal to D or is a refinement of D. In any event, we certainly have
$$\mathcal{U}(f, D') - \mathcal{L}(f, D') < \epsilon.$$

Let $D_1 = D' \cap [a, b]$, a dissection of $[a, b]$, and $D_2 = D' \cap [b, c]$, a dissection of $[b, c]$. If we write

$$D_1 = \{a = x_0, x_1, \ldots, x_m = b\}, \quad D_2 = \{b = x_m, x_{m+1}, \ldots, x_n = c\}, \tag{5.8}$$

then, with the usual notation,

$$0 \leq \mathcal{U}(f, D_1) - \mathcal{L}(f, D_1)$$
$$= \sum_{i=1}^m (M_i - m_i)(x_i - x_{i-1}) \leq \sum_{i=1}^n (M_i - m_i)(x_i - x_{i-1})$$
$$= \mathcal{U}(f, D') - \mathcal{L}(f, D') < \epsilon$$

and similarly $\mathcal{U}(f, D_2) - \mathcal{L}(f, D_2) < \epsilon$. Hence $f \in \mathcal{R}[a, b]$, $f \in \mathcal{R}[b, c]$.

Conversely, suppose that $f \in \mathcal{R}[a, b]$, $f \in \mathcal{R}[b, c]$. Then there exist dissections D_1 of $[a, b]$ and D_2 of $[b, c]$ such that

$$\mathcal{U}(f, D_1) - \mathcal{L}(f, D_1) < \frac{\epsilon}{2}, \quad \mathcal{U}(f, D_2) - \mathcal{L}(f, D_2) < \frac{\epsilon}{2}.$$

Let $D = D_1 \cup D_2$, a dissection of $[a, c]$. Then

$$\mathcal{U}(f, D) - \mathcal{L}(f, D) = [\mathcal{U}(f, D_1) + \mathcal{U}(f, D_2)] - [\mathcal{L}(f, D_1) + \mathcal{L}(f, D_2)]$$
$$= [\mathcal{U}(f, D_1) - \mathcal{L}(f, D_1)] + [\mathcal{U}(f, D_2) - \mathcal{L}(f, D_2)]$$
$$< \frac{\epsilon}{2} + \frac{\epsilon}{2} = \epsilon,$$

and so $f \in \mathcal{R}[a, c]$.

To show the equality (5.7), we use the same dissections D_1, D_2 and D as above. Then

$$0 \le \mathcal{U}(f, D) - \int_a^c f < \epsilon, \quad 0 \le \int_a^c f - \mathcal{L}(f, D) < \epsilon,$$

and so

$$\mathcal{U}(f, D) - \epsilon < \int_a^c f < \mathcal{L}(f, D) + \epsilon. \tag{5.9}$$

Similarly,

$$-\mathcal{L}(f, D_1) - \frac{\epsilon}{2} < -\int_a^b f < -\mathcal{U}(f, D_1) + \frac{\epsilon}{2} \tag{5.10}$$

$$-\mathcal{L}(f, D_2) - \frac{\epsilon}{2} < -\int_b^c f < -\mathcal{U}(f, D_2) + \frac{\epsilon}{2}. \tag{5.11}$$

By adding (5.10), (5.11) and (5.9) we obtain

$$\mathcal{U}(f, D) - \mathcal{L}(f, D) - 2\epsilon < \int_a^c f - \left(\int_a^b f + \int_b^c f \right) < \mathcal{L}(f, D) - \mathcal{U}(f, D) + 2\epsilon.$$

Hence certainly

$$-2\epsilon < \int_a^c f - \left(\int_a^b f + \int_b^c f \right) < 2\epsilon.$$

Since this holds for every $\epsilon > 0$, we are forced to conclude that $\int_a^c f = \int_a^b f + \int_b^c f$. $\qquad \square$

Remark 5.13

If we define, for $a < b$,

$$\int_a^a f = 0, \qquad \int_b^a f = -\int_a^b f,$$

we can extend the validity of the formula (5.10) to arbitrary a, b and c. For example, if $a \le c \le b$, then, by (5.10)

$$\int_a^b f = \int_a^c f + \int_c^b f = \int_a^c f - \int_b^c f,$$

and so

$$\int_a^c f = \int_a^b f + \int_b^c f \, ,$$

as required.

Given two functions f defined on $[a, b]$, it is reasonable to write $f \leq g$ if $f(x) \leq g(x)$ for all x in $[a, b]$. We then have the following result:

Theorem 5.14

If f, g in $\mathcal{R}[a, b]$ are such that $f \leq g$, then $\int_a^b f \leq \int_a^b g$.

Proof

Let $D = \{a = x_0, x_1, \ldots, x_n = b\}$ be a dissection of $[a, b]$. For $i = 1, 2, \ldots, n$, let $M_i = \sup_{(x_{i-1}, x_i)} f$, $P_i = \sup_{(x_{i-1}, x_i)} g$. Then $M_i \leq P_i$ for all i, and so

$$\mathcal{U}(f, D) = \sum_{i=1}^n M_i(x_i - x_{i-1}) \leq \sum_{i=1}^n P_i(x_i - x_{i-1}) = \mathcal{U}(g, D) \, .$$

This holds for every dissection D in $\mathcal{D}[a, b]$, and so

$$\int_a^b f = \inf_{D \in \mathcal{D}[a,b]} \mathcal{U}(f, D) \leq \inf_{D \in \mathcal{D}[a,b]} \mathcal{U}(g, D) = \int_a^b g \, .$$

$\square$

Theorem 5.15

Let $f, g \in \mathcal{R}[a, b]$. Then:

(i) $f + g \in \mathcal{R}[a, b]$, and

$$\int_a^b f + g = \int_a^b f + \int_a^b g \, ;$$

(ii) $kf \in \mathcal{R}[a, b]$ for every constant k, and

$$\int_a^b kf = k \int_a^b f \, ;$$

(iii) $f - g \in \mathcal{R}[a, b]$, and

$$\int_a^b f - g = \int_a^b f - \int_a^b g \, ;$$

(iv) $|f| \in \mathcal{R}[a, b]$, and

$$\left| \int_a^b f \right| \leq \int_a^b |f| \, ;$$

(v) $f \cdot g \in \mathcal{R}[a, b]$.

Proof

(i) Let $D = \{a = x_0, x_1, \ldots, x_n = b\}$ be a dissection of $[a, b]$ and, as usual, let us write $M_i = \sup_{(x_{i-1}, x_i)} f$ and $m_i = \inf_{(x_{i-1}, x_i)} f$. Let

$$P_i = \sup_{(x_{i-1}, x_i)} g \qquad p_i = \inf_{(x_{i-1}, x_i)} g$$
$$Q_i = \sup_{(x_{i-1}, x_i)} (f + g) \qquad q_i = \inf_{(x_{i-1}, x_i)} (f + g) \, .$$

Then

$$Q_i \leq M_i + P_i, \qquad q_i \geq m_i + p_i \, ,$$

and so

$$Q_i - q_i \leq (M_i + P_i) - (m_i + p_i) = (M_i - m_i) + (P_i - p_i) \, .$$

Let $\epsilon > 0$ be given, and choose D so that

$$\mathcal{U}(f, D) - \mathcal{L}(f, D) < \frac{\epsilon}{2}, \quad \mathcal{U}(g, D) - \mathcal{L}(g, D) < \frac{\epsilon}{2} \, .$$

Then

$$\mathcal{U}(f + g, D) - \mathcal{L}(f + g, D) = \sum_{i=1}^{n} (Q_i - q_i)(x_i - x_{i-1})$$

$$\leq \sum_{i=1}^{n} (M_i - m_i)(x_i - x_{i-1}) + \sum_{i=1}^{n} (P_i - p_i)(x_i - x_{i-1})$$

$$= [\mathcal{U}(f, D) - \mathcal{L}(f, D)] + [\mathcal{U}(g, D) - \mathcal{L}(g, D)]$$

$$< \frac{\epsilon}{2} + \frac{\epsilon}{2} = \epsilon \, ,$$

and so $f + g \in \mathcal{R}[a, b]$.

Suppose again that D is such that

$$\mathcal{U}(f, D) - \mathcal{L}(f, D) < \frac{\epsilon}{2}, \quad \mathcal{U}(g, D) - \mathcal{L}(g, D) < \frac{\epsilon}{2} \, .$$

Then, since the integral lies between the upper and lower sums, we can deduce that

$$\mathcal{L}(f, D) > \int_a^b f - \frac{\epsilon}{2}, \quad \mathcal{L}(g, D) > \int_a^b g - \frac{\epsilon}{2} \, ,$$

$$\mathcal{U}(f, D) < \int_a^b f + \frac{\epsilon}{2}, \quad \mathcal{U}(g, D) < \int_a^b g + \frac{\epsilon}{2} \, .$$

Then, from

$$\mathcal{U}(f + g, D) \leq \mathcal{U}(f, D) + \mathcal{U}(g, D)$$

we may deduce that

$$\int_a^b (f + g) \leq \mathcal{U}(f + g, D) \leq \mathcal{U}(f, D) + \mathcal{U}(g, D)$$

$$< \left(\int_a^b f + \frac{\epsilon}{2} \right) + \left(\int_a^b g + \frac{\epsilon}{2} \right) = \left(\int_a^b f + \int_a^b g \right) + \epsilon.$$

Similarly, from

$$\mathcal{L}(f + g, D) \geq \mathcal{L}(f, D) + \mathcal{L}(g, D)$$

we deduce that

$$\int_a^b (f + g) \geq \mathcal{L}(f + g, D) \geq \mathcal{L}(f, D) + \mathcal{L}(g, D)$$

$$> \left(\int_a^b f - \frac{\epsilon}{2} \right) + \left(\int_a^b g - \frac{\epsilon}{2} \right) = \left(\int_a^b f + \int_a^b g \right) - \epsilon.$$

Thus

$$\left| \int_a^b (f + g) - \left(\int_a^b f + \int_a^b g \right) \right| < \epsilon,$$

and, since this holds for *every* positive ϵ, we conclude that

$$\int_a^b (f + g) = \int_a^b f + \int_a^b g,$$

as required.

(ii) If $k = 0$ then it is clear that

$$\int_a^b kf = k \int_a^b f = 0.$$

Otherwise, let D, M_i and m_i be as before. Suppose first that $k > 0$. Then

$$\sup_{(x_{i-1}, x_i)} (kf) = kM_i, \qquad \inf_{(x_{i-1}, x_i)} (kf) = km_i$$

and so

$$\mathcal{U}(kf, D) = \sum_{i=1}^n (kM_i)(x_{i-1} - x_i) = k\mathcal{U}(f, D), \qquad (5.12)$$

$$\mathcal{L}(kf, D) = \sum_{i=1}^n (km_i)(x_{i-1} - x_i) = k\mathcal{L}(f, D).$$

Since $f \in \mathcal{R}[a, b]$, we can choose D so that, for a given $\epsilon > 0$,

$$\mathcal{U}(f, D) - \mathcal{L}(f, D) < \epsilon/k,$$

and it then follows that $\mathcal{U}(kf, D) - \mathcal{L}(kf, D) < \epsilon$. Thus $kf \in \mathcal{R}[a, b]$.

Next, from (5.12) we deduce that

$$\int_a^b (kf) = \inf_{D \in \mathcal{D}[a,b]} \mathcal{U}(kf, D) = k \inf_{D \in \mathcal{D}[a,b]} \mathcal{U}(f, D) = k \int_a^b f.$$

Next, suppose that $k < 0$, say $k = -l$, where $l > 0$. Then, from (i),

$$0 = \int_a^b ((-l)f + lf) = \int_a^b (-l)f + \int_a^b lf = \int_a^b (-l)f + l \int_a^b f,$$

and so

$$\int_a^b kf = \int_a^b (-l)f = (-l) \int_a^b f = k \int_a^b f.$$

(iii) This follows immediately from (i) and (ii), since $f - g = f + (-1)g$.

(iv) We show first that $|f| \in \mathcal{R}[a, b]$. Let D, M_i and m_i be defined as before, and, for each i in $\{1, 2, \ldots, n\}$, let us write

$$A_i = \sup_{(x_{i-1}, x_i)} |f|, \quad a_i = \inf_{(x_{i-1}, x_i)} |f|.$$

For all x, y in (x_{i-1}, x_i), we know that

$$|f|(x) - |f|(y) = |f(x)| - |f(y)| \leq |f(x) - f(y)| \leq M_i - m_i.$$

Thus

$$|f|(x) \leq M_i - m_i + |f|(y) \text{ for all } x \text{ in } (x_{i-1}, x_i).$$

It follows that

$$A_i \leq M_i - m_i + |f|(y),$$

and so

$$|f|(y) \geq A_i - M_i + m_i \text{ for all } y \text{ in } (x_{i-1}, x_i).$$

Hence $a_i \geq A_i - M_i + m_i$, and so

$$A_i - a_i \leq M_i - m_i.$$

Now let $\epsilon > 0$ be given, and let D be such that

$$\mathcal{U}(f, D) - \mathcal{L}(f, D) = \sum_{i=1}^n (M_i - m_i)(x_i - x_{i-1}) < \epsilon.$$

Then

$$\mathcal{U}(|f|, D) - \mathcal{L}(|f|, D) = \sum_{i=1}^{n} (A_i - a_i)(x_i - x_{i-1})$$

$$\leq \sum_{i=1}^{n} (M_i - m_i)(x_i - x_{i-1}) < \epsilon,$$

and so $|f| \in \mathcal{R}[a, b]$.

The inequality follows easily from Theorem 5.14. We have

$$f \leq |f|, \qquad -f = (-1)f \leq |f|,$$

and so

$$\int_a^b f \leq \int_a^b |f|, \qquad -\int_a^b f = \int_a^b (-f) \leq \int_a^b |f|.$$

Hence

$$\left| \int_a^b f \right| \leq \int_a^b |f|,$$

as required.

(v) We show first that if $f \in \mathcal{R}[a, b]$ then $f \cdot f$, which for the moment[1] we can reasonably write as f^2, is also in $\mathcal{R}[a, b]$. Since $f^2 = |f|^2$, we may assume that $f(x) \geq 0$ for all x in $[a, b]$.

Let D, M_i, m_i be as before, and let $M = \sup_{[a,b]} f$, $m = \inf_{[a,b]} f$. If $f(x) = 0$ for all x in $[a, b]$ then there is nothing to prove. We may therefore suppose that $M > 0$. Let $P_i = \sup_{(x_{i-1}, x_i)} (f^2)$, $p_i = \inf_{(x_{i-1}, x_i)} (f^2)$. Then $P_i = M_i^2$, $p_i = m_i^2$, and so

$$P_i - p_i = (M_i + m_i)(M_i - m_i) \leq 2M(M_i - m_i).$$

For each $\epsilon > 0$ we can choose D so that

$$\mathcal{U}(f, D) - \mathcal{L}(f, D) < \frac{\epsilon}{2M}.$$

Then

$$\mathcal{U}(f^2, D) - \mathcal{L}(f^2, D) = \sum_{i=1}^{n} (P_i - p_i)(x_i - x_{i-1})$$

$$\leq 2M \sum_{i=1}^{n} (M_i - m_i)(x_i - x_{i-1})$$

$$= 2M \left(\mathcal{U}(f^2, D) - \mathcal{L}(f^2, D) \right)$$

$$< \epsilon,$$

---

[1] We do have to be careful in using the notation f^2, which might mean either $f \cdot f$ or $f \circ f$.

and so $f^2 \in \mathcal{R}[a, b]$.

The proof is completed with the observation that

$$f \cdot g = \frac{(f + g)^2 - (f - g)^2}{4} \, .$$

From (ii) and (iii) we see that $f + g, f - g \in \mathcal{R}[a, b]$, then by what we have just shown it follows that $(f + g)^2, (f - g)^2 \in \mathcal{R}[a, b]$, and finally the required result follows from (i) and (ii).

$\square$

EXERCISES

5.10 Give an example of functions f, g in $\mathcal{R}[a, b]$ such that

$$\int_a^b f \cdot g \neq \left(\int_a^b f \right) \left(\int_a^b g \right) .$$

5.11 For continuous functions f, g on $[a, b]$, define the **inner product** (f, g) to be $\int_a^b f \cdot g$. Show that, for all f, g, h and all constants k,

a) $(f + g, h) = (f, h) + (g, h)$, $(f, g + h) = (f, g) + (f, h)$;

b) $(kf, g) = k(f, g)$, $(f, kg) = k(f, g)$;

c) $(f, f) \geq 0$, and $(f, f) = 0$ only if $f = 0$.

Establish the following version of the **Cauchy–Schwarz inequality**:

$$\left(\int_a^b f \cdot g \right)^2 \leq \left(\int_a^b f^2 \right) \left(\int_a^b g^2 \right) .$$

5.4 The Fundamental Theorem

In this section we establish the crucial connection between integration and differentiation. We begin with the following easy result:

Theorem 5.16

Let $f \in \mathcal{R}[a, b]$ and let $M = \sup_{[a,b]} f$, $m = \inf_{[a,b]} f$. Then

$$m(b - a) \leq \int_a^b f \leq M(b - a).$$

Proof

Geometrically this is fairly clear, since the area under the curve $y = f(x)$ is between the areas $M(b - a)$ and $m(b - a)$ of the upper and lower rectangles.

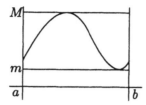

Analytically, we can argue as follows. Consider the dissection $D = \{a, b\}$ consisting solely of the two endpoints. Then

$$m(a - b) = \mathcal{L}(f, D) \le \int_a^b f \le \mathcal{U}(f, D) = M(b - a).$$

$\square$

If the function f is continuous, we can use this result to prove what is sometimes called the **Mean Value Theorem of Integral Calculus**:

Theorem 5.17 (The Mean Value Theorem)

Let f be continuous in $[a, b]$. Then there exists c in $[a, b]$ such that

$$\int_a^b f = (b - a)f(c).$$

Proof

From Theorem 5.16 we know that

$$m \le \frac{1}{b - a} \int_a^b f \le M.$$

By the Intermediate Value Theorem (Theorem 3.12), we deduce that there exists c in $[a, b]$ such that

$$f(c) = \frac{1}{b - a} \int_a^b f,$$

and the result follows.

$\square$

The next, crucially important result now follows easily:

Theorem 5.18 (The Fundamental Theorem of Calculus)

Let f be continuous in $[a, b]$. For each x in $[a, b]$ define

$$F(x) = \int_a^x f \,.$$

Then F is differentiable in $[a, b]$, and $F'(x) = f(x)$.

Proof

From Theorem 5.12 we have that, for all x, $x + h$ in $[a, b]$,

$$F(x + h) - F(x) = \int_x^{x+h} f \,.$$

(In view of Remark 5.13 this holds good even if h is negative.) Hence, by Theorem 5.17,

$$F(x + h) - F(x) = h f(x + \theta h) \,,$$

where $0 \le \theta \le 1$. As $h \to 0$ we know that $f(x + \theta h) \to f(x)$, since f is continuous. Hence $F'(x)$ exists, and, for all x in $[a, b]$,

$$F'(x) = f(x) \,.$$

$\square$

Remark 5.19

When we say that F is differentiable in the closed interval $[a, b]$, we mean that F has a derivative for every x in the open interval (a, b), a right derivative at a and a left derivative at b.

This theorem is the key to the most important standard technique in integration. Given a function f, continuous on $[a, b]$, let Φ be an "antiderivative" of f, that is to say, a function such that $\Phi' = f$. Then if, as above, we define

$$F(x) = \int_a^x f \quad (x \in [a, b]) \,,$$

we have that $F'(x) = f(x) = \Phi'(x)$. It follows from Theorem 4.8 that, for all x in $[a, b]$,

$$\int_a^x f = F(x) = \Phi(x) + C$$

for some constant C. Putting $x = a$ gives

$$\Phi(a) + C = F(a) = \int_a^a f = 0$$

and so $C = -\Phi(a)$. Then putting $x = b$ gives:

Theorem 5.20

Let f be a continuous function on $[a, b]$, and let $\Phi' = f$. Then

$$\int_a^b f = \Phi(b) - \Phi(a).$$

$\square$

So, for example, if $f(x) = x^3$, we may take

$$\Phi(x) = \frac{1}{4} x^4$$

and obtain

$$\int_1^2 f = \Phi(2) - \Phi(1) = \frac{1}{4}(16 - 1) = \frac{15}{4}.$$

We usually write this as

$$\int_1^2 x^3 \, dx = \left[\frac{x^4}{4} \right]_1^2 = \frac{1}{4}(16 - 1) = \frac{15}{4}.$$

Remark 5.21

It is important to note that, while for every f in $C[a, b]$ a suitable Φ does always exist – simply choose $\Phi(x) = \int_a^x f(t) \, dt$ – it is not always possible to express Φ in terms of "known" functions.

One aspect of Theorem 5.18 to which we have not yet drawn attention is that the integration process "improves" the function: f is merely continuous, while F has the stronger property of being differentiable. The following theorem, with which we close this section, is also concerned with improvement.

Theorem 5.22

Let $f \in \mathcal{R}[a, b]$ and let

$$F(x) = \int_a^x f \quad (x \in [a, b]).$$

Then F is continuous in $[a, b]$.

Proof

Let $c \in [a, b]$, and let $\epsilon > 0$. For all x in $[a, b]$ we have (allowing for the possibility that $x < c$)

$$|F(x) - F(c)| = \left| \int_c^x f \right| \leq \left| \int_c^x |f| \right|,$$

by virtue of Theorem 5.15(iv). Let $K = \sup_{[a,b]} |f|$. Then, by Theorem 5.16,

$$\int_c^x |f| \leq K|x - c|,$$

and so $|F(x) - F(c)| \leq K|x - c|$. If we now choose $\delta = \epsilon/K$ we can assert that $|F(x) - F(c)| < \epsilon$ whenever x, in $[a, b]$, is such that $|x - c| < \delta$. Thus F is continuous, as required.   □

EXERCISES

5.12 Show that the mean value theorem (Theorem 5.17) may fail if $f \in \mathcal{R}[a, b]$ but is not continuous.

5.13 Prove the following generalization of the mean value theorem.

Let f be continuous on $[a, b]$, and let g be a function such that $g(x) \geq 0$ for all x in $[a, b]$ and $\int_a^b g > 0$. Then there exists c in $[a, b]$ such that

$$\int_a^b f \cdot g = f(c) \int_a^b g.$$

5.14 Let g be continuous, and let

$$f(x) = \frac{1}{2} \int_0^x (x - t)^2 g(t) \, dt.$$

Show that

$$f'(x) = x \int_0^x g(t) \, dt - \int_0^x t g(t) \, dt,$$

and find $f''(x)$, $f'''(x)$.

5.15 Let

$$f(x) = \int_{x^3}^{x^2} \frac{t^5 \, dt}{1 + t^4}.$$

Without attempting to evaluate the integral, find $f'(x)$.

5.5 Techniques of Integration

It is traditional to refer to antiderivatives as **indefinite integrals,** and to write

$$F(x) = \int f(x)\,dx$$

if $F'(x) = f(x)$. By contrast, we refer to $\int_a^b f$, often written as $\int_a^b f(x)\,dx$, as a **definite integral.** Note that, by virtue of Theorem 5.20, if $F(x) = \int f(x)\,dx$, then

$$\int_a^b f = F(b) - F(a)\,.$$

From our knowledge of differentiation, certain indefinite integrals are obvious:

$$\int x^n\,dx = \frac{x^{n+1}}{n+1} \quad (n \neq -1)\,, \tag{5.13}$$

$$\int \sin x\,dx = -\cos x\,, \quad \int \cos x\,dx = \sin x\,, \tag{5.14}$$

$$\int \frac{dx}{\sqrt{1-x^2}} = \sin^{-1} x\,, \quad \int \frac{dx}{1+x^2} = \tan^{-1} x\,. \tag{5.15}$$

From Theorem 4.2 we know that $k \int f(x)\,dx + l \int g(x)\,dx$ is an indefinite integral of $kf + lg$. Hence polynomial functions can be integrated in a straightforward way: for example,

$$\int (x^3 + 7x - 2)\,dx = \frac{x^4}{4} + \frac{7x^2}{2} - 2x\,.$$

It is important to realise that an indefinite integral really is indefinite: it is always possible to add an arbitrary constant to the answer. Thus both the statements

$$\int (x+1)^2\,dx = \int (x^2 + 2x + 1)\,dx = \frac{x^3}{3} + x^2 + x \tag{5.16}$$

and

$$\int (x+1)^2\,dx = \frac{(x+1)^3}{3} \tag{5.17}$$

are correct, although the two answers are not identical. For this reason, there is a common convention, whenever $\Phi' = f$, of writing

$$\int f(x) = \Phi(x) + C\,,$$

where C is an "arbitrary constant".

Notice that both (5.16) and (5.17) lead to the correct answer if we insert limits to obtain a definite integral:

$$\int_0^1 (x+1)^2 = \left[\frac{x^3}{3} + x^2 + x\right]_0^1 = \frac{1}{3} + 1 + 1 = \frac{7}{3},$$

$$\int_0^1 (x+1)^2 = \left[\frac{(x+1)^3}{3}\right]_0^1 = \frac{8}{3} - \frac{1}{3} = \frac{7}{3}.$$

Theorem 5.15 tells us that the product of two integrable functions is integrable, but conspicuously fails to give any indication of the the relationship between $\int_a^b f \cdot g$ on the one hand, and $\int_a^b f$ and $\int_a^b g$ on the other. If we hope to clarify this relationship, it is natural to look to the product formula for differentiation. From Theorem 4.2 we know that, for differentiable functions f, h defined on a suitable domain

$$(f \cdot h)'(x) = f(x)h'(x) + f'(x)h(x),$$

and from this it follows, if f' and h' are continuous, that

$$f(x)h(x) = \int (f \cdot h)'(x)\,dx = \int f(x)h'(x)\,dx + \int f'(x)h(x)\,dx.$$

If we now write $h'(x)$ as $g(x)$, so that $h(x) = \int g(x)\,dx$, we obtain

$$f(x)\int g(x)\,dx = \int f(x)g(x)\,dx + \int f'(x)\left[\int g(x)\,dx\right]dx.$$

This gives the formula for **integration by parts**, valid when f' and g are continuous:

$$\int f(x)g(x)\,dx = f(x)\int g(x)\,dx - \int f'(x)\left[\int g(x)\,dx\right]dx. \qquad (5.18)$$

If, on the right, we revert to writing $\int g(x)\,dx$ as $h(x)$, we can interpret the formula as

$$\int f(x)g(x)\,dx = f(x)h(x) - \int f'(x)h(x)\,dx.$$

Putting first $x = b$, then $x = a$, and then subtracting, we obtain

$$\int_a^b f(x)g(x)\,dx = \Big[f(x)h(x)\Big]_a^b - \int_a^b f'(x)h(x)\,dx. \qquad (5.19)$$

This is by no means a universal recipe for finding the integral of a product, but it does sometimes work. In particular, it can be useful if the function f is a polynomial, so that it disappears to 0 after enough applications of differentiation.

Example 5.23

Find $\int_0^{\pi/2} x^2 \sin x \, dx$.

Solution

Choose $f(x) = x^2$, $g(x) = \sin x$. From (5.14) we know that $h(x) = -\cos x$. So from (5.19) we have

$$\int_0^{\pi/2} x^2 \sin x \, dx = \left[x^2(-\cos x) \right]_0^{\pi/2} - \int_0^{\pi/2} 2x(-\cos x) \, dx \, .$$

Now $\cos(\pi/2) = 0$, and so the first term on the right vanishes. Thus

$$\int_0^{\pi/2} x^2 \sin x \, dx = 2 \int_0^{\pi/2} x \cos x \, dx \, . \qquad (5.20)$$

We use (5.19) again, this time with $f(x) = x$, $g(x) = \cos x$ and $h(x) = \int \cos x \, dx = \sin x$. Then

$$\int_0^{\pi/2} x \cos x \, dx = \left[x \sin x \right]_0^{\pi/2} - \int_0^{\pi/2} 1 . \sin x \, dx$$

$$= \frac{\pi}{2} - \left[-\cos x \right]_0^{\pi/2}$$

$$= \frac{\pi}{2} - [0 - (-1)] = \frac{\pi}{2} - 1 \, .$$

From (5.20) we now conclude that

$$\int_0^{\pi/2} x^2 \sin x \, dx = \pi - 2 \, .$$

$\square$

The technique of this example will work, though the details are tedious, for $x^3 \sin x$, $x^4 \sin x$, etc., but (for example) is of no use for $\sqrt{x} \sin x$.

Integration by parts can be useful in some slightly unexpected ways:

Example 5.24

Determine I_n, where

$$I_n = \int_0^{\pi/2} \sin^n x \, dx \quad (n = 0, 1, 2, \ldots) \, .$$

Solution

It is clear that

$$I_0 = \int_0^{\pi/2} 1 \, dx = \left[x\right]_0^{\pi/2} = \frac{\pi}{2}, \quad I_1 = \int_0^{\pi/2} \sin x \, dx = \left[-\cos x\right]_0^{\pi/2} = 1.$$

Let $n \geq 2$. In (5.19) let $f(x) = \sin^{n-1} x$ and $g(x) = \sin x$. Then $h(x) = \int g(x) \, dx = -\cos x$ and $f'(x) = (n-1) \sin^{n-2} x \cdot \cos x$. Hence

$$I_n = \left[\sin^{n-1} x(-\cos x)\right]_0^{\pi/2} - \int_0^{\pi/2} (n-1) \sin^{n-2} x \cdot \cos x(-\cos x) \, dx$$

$$= 0 + (n-1) \int_0^{\pi/2} (n-1) \sin^{n-2} x \cdot (1 - \sin^2 x), dx$$

$$= (n-1)(I_{n-2} - I_n).$$

Hence

$$n I_n = (n-1) I_{n-2},$$

and so we obtain the **reduction formula**

$$I_n = \frac{n-1}{n} I_{n-2}.$$

Using this formula repeatedly, we find that, for even n,

$$I_n = \frac{n-1}{n} \cdot \frac{n-3}{n-2} \cdots \cdots \frac{1}{2} \cdot \frac{\pi}{2}, \tag{5.21}$$

and, for odd n

$$I_n = \frac{n-1}{n} \cdot \frac{n-3}{n-2} \cdots \cdots \frac{2}{3}. \tag{5.22}$$

□

We arrived at the formula for integration by parts by considering the consequences for integration theory of the product rule for differentiation. In the same spirit we now consider the formula

$$(p \circ q)'(x) = p'\big(q(x)\big) q'(x)$$

(see Theorem 4.4). If we integrate both sides of this we obtain

$$p(q(x)) = \int p'\big(q(x)\big) q'(x) \, dx,$$

and if we tweak the notation a bit we arrive at the very useful technique of **integration by substitution**. First, we replace p' by f and p by F, so that $F' = f$:

$$\int f(q(x)) q'(x) \, dx = F(q(x)). \tag{5.23}$$

For example,

$$\int x^2 \sqrt{1+x^3} \, dx = \frac{1}{3} \int f(q(x)) q'(x) \, dx \,,$$

where $f(u) = \sqrt{u}$ and $q(x) = 1+x^3$. Now $F(u) = (2/3)u^{3/2}$, and so from (5.23) we deduce that

$$\int x^2 \sqrt{1+x^3} \, dx = \frac{2}{9}(1+x^3)^{3/2} \,.$$

In order to apply this idea correctly to definite integrals. we must ensure that the differentiable "substitution function" q has an inverse, which in effect means that its derivative must either be positive throughout, or negative throughout, the relevant range. The following theorem is certainly not the most general result possible, but it will suffice for the applications that we intend to make:

Theorem 5.25

Let f be continuous in $[a, b]$, and let g be a function whose derivative g' is positive throughout in an interval containing $[a, b]$. Then

$$\int_a^b f(x) \, dx = \int_{g(a)}^{g(b)} f\big(g^{-1}(u)\big) . (g^{-1})'(u) \, du \,. \tag{5.24}$$

Proof

From Theorem 4.15 we know that g has an inverse, and that its inverse has a positive derivative in $[g(a), g(b)]$. If F is chosen so that $F' = f$, then

$$\int_{g(a)}^{g(b)} f\big(g^{-1}(u)\big) . (g^{-1})'(u) \, du = \int_{g(a)}^{g(b)} F'\big(g^{-1}(u)\big) . (g^{-1})'(u) \, du$$

$$= \int_{g(a)}^{g(b)} (F \circ g^{-1})'(u) \, du = \Big[(F \circ g^{-1})(u) \Big]_{g(a)}^{g(b)}$$

$$= F(b) - F(a) = \int_a^b f(x) \, dx \,.$$

$\square$

The appearance of formula (5.24) suggests that we have replaced something simple by something complicated. The key, however, lies in the choice of the substitution function g.

Example 5.26

Evaluate

$$I = \int_{-1/2}^{1/2} \sqrt{1 - x^2}\, dx\,.$$

Solution

Let $g(x) = \sin^{-1} x$. Then g has a continuous positive derivative $1/\sqrt{1-x^2}$ throughout the interval $[-1/2, 1/2]$. We have

$$g\left(-\frac{1}{2}\right) = -\frac{\pi}{6}\,,\; g\left(\frac{1}{2}\right) = \frac{\pi}{6}\,,\; g^{-1}(u) = \sin u\,,\; (g^{-1})'(u) = \cos u\,.$$

Hence

$$I = \int_{-\pi/6}^{\pi/6} \sqrt{1 - \sin^2 u}\, \cos u\, du = \int_{-\pi/6}^{\pi/6} \cos^2 u\, du\,,$$

since $\cos u$ is positive throughout $[-\pi/6, \pi/6]$. From Exercise 3.10 we deduce that

$$I = \frac{1}{2} \int_{-\pi/6}^{\pi/6} (1 + \cos 2u)\, du\,.$$

Finally, noting that the derivative of $\sin 2u$ is $2\cos 2u$, we obtain

$$I = \frac{1}{2} \left[u + \frac{1}{2} \sin 2u \right]_{-\pi/6}^{\pi/6}$$

$$= \frac{1}{2} \left[\frac{\pi}{6} + \frac{\sqrt{3}}{4} - \left(-\frac{\pi}{6}\right) - \left(-\frac{\sqrt{3}}{4}\right) \right]$$

$$= \frac{\pi}{6} + \frac{\sqrt{3}}{4}\,.$$

$\square$

This is what lies behind the standard "calculus" approach, which involves an argument something like this. Let $x = \sin u$. Then $x = \pm 1/2$ gives $u = \pm \pi/6$, and $dx/du = \cos u$, giving $dx = \cos u\, du$. Thus

$$I = \int_{-\pi/6}^{\pi/6} \cos^2 u\, du\,,$$

just as before.

If the substitution function has *negative* derivative throughout, there is no great problem:

Example 5.27

Evaluate

$$I = \int_0^{1/2} \frac{x\,dx}{(1-x^2)^2}.$$

Solution

Let $g(x) = 1 - x^2$. Then $g(0) = 0$, $g(1/2) = 3/4$,

$$g^{-1}(u) = (1-u)^{1/2}, \quad (g^{-1})'(u) = -\frac{1}{2}(1-u)^{-1/2},$$

and so

$$I = \int_1^{3/4} (1-u)^{1/2} \cdot \frac{1}{u^2} \cdot \left(-\frac{1}{2}(1-u)^{-1/2}\right) du = \frac{1}{2} \int_{3/4}^1 \frac{du}{u^2}$$
$$= \frac{1}{2}\left[-\frac{1}{u}\right]_{3/4}^1 = \frac{1}{2}\left(-1 + \frac{4}{3}\right) = \frac{1}{6}.$$

In practice we can avoid a lot of the calculation. Let $u = 1 - x^2$. Then $du = -2x\,dx$, and so $x\,dx = (-1/2)du$. Thus

$$I = \frac{1}{2} \int_1^{3/4} \frac{-du}{u^2} = \frac{1}{2} \int_{3/4}^1 \frac{du}{u^2}.$$

$\square$

EXERCISES

5.16 Evaluate

$$\int_0^{\pi/4} \tan^2 x\,dx, \quad \int_0^{\pi/2} \sin^2 x\,dx.$$

5.17 What is wrong with the following argument?

$$\int \frac{dx}{x} = \int u(x)v(x)\,dx, \quad \text{where } u(x) = x,\ v(x) = 1/x^2$$
$$= u(x) \int v(x)\,dx - \int u'(x)\left[\int v(x)\,dx\right] dx$$
$$= x\left(-\frac{1}{x}\right) - \int 1\left(-\frac{1}{x}\right) dx$$
$$= -1 + \int \frac{dx}{x},$$

and so $0 = -1$.

5.18 Let

$$I_n = \int_0^{\pi/2} x \sin^n x \, dx \, .$$

Evaluate I_0 and I_1, and show that, for all $n \geq 2$,

$$I_n = \frac{1}{n^2} + \frac{n-1}{n} I_{n-2} \, .$$

Hence evaluate I_3 and I_4.

5.19 Evaluate

$$\int_0^1 \frac{x^5 \, dx}{\sqrt{1+x^6}} \, , \quad \int_0^{\pi/2} \frac{\sin x \, dx}{(3 + \cos x)^2} \, , \quad \int_0^{\pi/4} \cos 2x \sqrt{4 - \sin 2x} \, dx \, .$$

5.20 Use the substitution $u = \pi - x$ to show that, for every suitable function f,

$$\int_0^\pi x f(\sin x) \, dx = \frac{\pi}{2} \int_0^\pi f(\sin x) \, dx \, .$$

Hence show that

$$\int_0^\pi \frac{x \sin x \, dx}{1 + \cos^2 x} = \frac{\pi^2}{4} \, .$$

5.21 Prove, by induction on n, the following alternative version of Taylor's Theorem:

Let f have a continuous nth derivative in some interval I containing a. Then, for every x in I

$$f(x) = \sum_{r=0}^{n-1} \frac{1}{r!} f^{(r)}(a)(x-a)^r + R_n \, ,$$

where

$$R_n = \frac{1}{(n-1)!} \int_a^x (x-t)^{n-1} f^{(n)}(t) \, dt \, .$$

5.6 Improper Integrals of the First Kind

So far, our theory of integration has dealt with bounded functions defined on finite intervals. For many applications we need to relax one or both of these conditions.

If we allow infinite intervals we obtain what we call an **improper integral of the first kind**, or, more simply, an **infinite integral**. If, for a fixed a,

$\lim_{K \to \infty} \int_a^K f$ exists, it is reasonable to denote the limit by $\int_a^\infty f$. Similarly, if $\lim_{L \to \infty} \int_{-L}^a f$ exists, we denote it by $\int_{-\infty}^a f$. If both limits exist, then we define

$$\int_{-\infty}^\infty f \quad \text{as} \quad \lim_{L \to \infty} \int_{-L}^a f + \lim_{K \to \infty} \int_a^K f .$$

(The value of $\int_{-\infty}^\infty f$ is of course independent of a.)

Remark 5.28

The existence of $\int_{-\infty}^\infty f$ requires the *separate* existence of $\int_a^\infty f$ and $\int_{-\infty}^a f$. This is a stronger requirement than the existence of $\lim_{K \to \infty} \int_{-K}^K f$. For example, consider the case of $f(x) = x$.

The theory of infinite integrals mirrors the theory of infinite series. As with infinite series, if $\int_a^\infty f$ exists, we say that the integral is **convergent**. Otherwise, we say that it is **divergent**

The following theorem is closely comparable to Theorem 2.25:

Theorem 5.29

The integral

$$\int_1^\infty \frac{dx}{x^n}$$

is convergent if and only if $n > 1$.

Proof

If $n \neq 1$ then

$$\int_1^K \frac{dx}{x^n} = \int_1^K x^{-n}\, dx = \left[\frac{x^{-n+1}}{-n+1} \right]_1^K = \frac{1}{n-1} \left[1 - K^{1-n} \right] .$$

If $n > 1$ then $1 - n < 0$, and so $K^{1-n} \to 0$ as $K \to \infty$. In this case the integral converges, and

$$\int_1^\infty \frac{dx}{x^n} = \frac{1}{n-1} .$$

If $n < 1$ then $1 - n > 0$, and so $K^{1-n} \to \infty$ as $K \to \infty$. In this case the integral diverges.

It remains to consider the case $n = 1$, where the formula for integrating $1/x^n$ does not apply. For every integer $N > 1$ denote $\int_1^N (dx/x)$ by I_N. Then

$$I_N = \int_1^2 \frac{dx}{x} + \int_2^3 \frac{dx}{x} + \cdots + \int_{N-1}^N \frac{dx}{x}. \tag{5.25}$$

Now, in the interval $[i-1, i]$ $(i = 2, 3, \ldots, N)$, we have that

$$\frac{1}{i} \leq \frac{1}{x} \leq \frac{1}{i-1}.$$

Hence, by Theorem 5.16,

$$\frac{1}{i} \leq \int_{i-1}^i \frac{dx}{x} \leq \frac{1}{i-1}.$$

It now follows from (5.25) that

$$I_N \geq \frac{1}{2} + \frac{1}{3} + \cdots + \frac{1}{N}.$$

Since the harmonic series diverges, we conclude that I_N can be made arbitrarily large by taking N sufficiently large. Thus

$$\int_1^\infty \frac{dx}{x}$$

is divergent, as required. $\square$

As with infinite series, it pays to concentrate first on the case when the **integrand** function f is non-negative. In this case the integral $\int_a^K f$ (for a fixed a) is an increasing function of K, and we can appeal to the following theorem. The proof is closely comparable to that of Theorem 2.9, and is left as an exercise.

Theorem 5.30

Let F be an increasing function, defined on the interval $[a, \infty)$. If F is bounded, then $\lim_{x \to \infty} F(x)$ exists.

We then have the simplest version of the **comparison test for infinite integrals**, closely comparable to the comparison test (Theorem 2.22) for series of positive terms:

Theorem 5.31 (The Comparison Test)

Let f, g be positive functions, integrable in every finite interval contained in $[a, \infty)$.

(i) If $\int_a^\infty f$ is convergent and $g \leq f$, then $\int_a^\infty g$ is convergent.

(ii) If $\int_a^\infty f$ is divergent and $g \geq f$, then $\int_a^\infty g$ is divergent.

Proof

(i) For all $K > a$ the function $G(K) = \int_a^K g$ has the property that

$$G(K) \leq \int_a^K f \leq \int_a^\infty f \,.$$

Hence, by Theorem 5.30, $\lim_{K \to \infty} G(K)$ exists. That is, $\int_a^\infty g$ is convergent.

(ii) This is clear, since if $\int_a^\infty g$ were convergent, then $\int_a^\infty f$ would be convergent by part (i). □

As with series, the comparison need not be valid for all x in $[a, \infty)$, but only for "sufficiently large" x. Also, the introduction of positive constants into the comparison makes no essential difference:

Theorem 5.32

Let f, g be positive functions, integrable in every finite interval contained in $[a, \infty)$. Let k be a positive constant.

(i) If $\int_a^\infty f$ is convergent and if there exists $M > a$ such that $g(x) \leq kf(x)$ for all $x > M$, then $\int_a^\infty g$ is convergent.

(ii) If $\int_a^\infty f$ is divergent and if there exists $M > a$ such that $g(x) \geq kf(x)$ for all $x > M$, then $\int_a^\infty g$ is divergent.

As with series, for two positive functions f and g defined on $[a, \infty)$ we can define $f \asymp g$ (as $x \to \infty$) to mean that

$$\lim_{x \to \infty} \frac{f(x)}{g(x)} = K \,,$$

for some $K > 0$. The stronger statement $f \sim g$ means that

$$\lim_{x \to \infty} \frac{f(x)}{g(x)} = 1 \,.$$

We can also define $f \asymp g$ (as $x \to -\infty$) and $f \sim g$ (as $x \to -\infty$) in the obvious way.

We then have an alternative version of the comparison test, which is often more convenient to use:

Theorem 5.33

Let f, g be positive bounded functions on $[a, \infty)$. If $f \asymp g$, then $\int_a^\infty f$ is convergent if and only if $\int_a^\infty g$ convergent.

Proof

Suppose that $\int_a^\infty f$ is convergent, with $\int_a^\infty f = I$. Suppose also that $f \asymp g$, so that there exists $K > 0$ such that

$$\lim_{x \to \infty} \frac{f(x)}{g(x)} = K.$$

Choosing $\epsilon = K/2$, we can assert that there exists M such that, for all $x > M$,

$$\left| \frac{f(x)}{g(x)} - K \right| < \frac{K}{2};$$

hence

$$\frac{K}{2} < \frac{f(x)}{g(x)} < \frac{3K}{2},\qquad(5.26)$$

and so $g(x) < (2/K)f(x)$ for all $x > M$. Hence, by Theorem 5.32, $\int_a^\infty g$ is convergent.

Conversely, if we suppose that $\int_a^\infty g$ is convergent, then from (5.26) we deduce that $f(x) < (3K/2)g(x)$ for all $x > M$, and again the convergence of $\int_a^\infty f$ follows from Theorem 5.32. $\qquad\square$

Example 5.34

Investigate the convergence of

$$\text{(a)} \int_0^\infty \frac{dx}{\sqrt{1 + x^2}}; \qquad \text{(b)} \int_0^\infty \frac{dx}{(1 + x^2)^{3/2}}.$$

Solution
(a) Since

$$\frac{\sqrt{1 + x^2}}{x} = \sqrt{1 + (1/x^2)} \to 1$$

as $x \to \infty$, we have

$$\frac{1}{\sqrt{1 + x^2}} \sim \frac{1}{x}.$$

Hence, by Theorems 5.29 and 5.31, the integral is divergent.

(b) Since

$$\frac{(1+x^2)^{3/2}}{x^3} = \left(1 + \frac{1}{x^2}\right)^{3/2} \to 1$$

as $x \to \infty$, we have

$$\frac{1}{(1+x^2)^{3/2}} \sim \frac{1}{x^3}.$$

Hence, again by Theorems 5.29 and 5.31, the integral is convergent. □

When it comes to the more difficult question concerning functions taking positive and negative values, we begin, as with series, by introducing the notion of absolute convergence: the integral $\int_a^\infty f$ is said to be **absolutely convergent** if $\int_a^\infty |f|$ is convergent. By analogy with Theorem 2.36, we have:

Theorem 5.35

Let f be a continuous function defined on the interval $[a, \infty)$, and such that $\int_a^\infty f$ is absolutely convergent. Then $\int_a^\infty f$ is convergent.

Proof

For x in $[a, \infty)$ let

$$g(x) = \begin{cases} f(x) & \text{when } f(x) \geq 0 \\ 0 & \text{when } f(x) < 0, \end{cases}$$

$$h(x) = \begin{cases} 0 & \text{when } f(x) \geq 0 \\ -f(x) & \text{when } f(x) < 0. \end{cases}$$

Then

$$f = g - h, \; |f| = g + h, \; g = \frac{1}{2}(|f| + f), \; h = \frac{1}{2}(|f| - f).$$

Now, $|f|$ is continuous by Theorem 3.11, and so, by the same theorem, g and h are both continuous. Thus, by Theorem 5.9, g and h are integrable in every finite interval contained in $[a, \infty)$. Since g and h are both positive functions and since, for all x in $[a, \infty)$,

$$g(x) \leq |f|(x), \quad h(x) \leq |f|(x),$$

it follows by the comparison test (Theorem 5.31) that $\int_a^\infty g$ and $\int_a^\infty h$ both converge. Now, for all $K > a$,

$$\int_a^K f = \int_a^K g - \int_a^K h.$$

Hence, taking limits as $K \to \infty$, we see that $\int_a^\infty f$ is convergent, with value $\int_a^\infty g - \int_a^\infty h$. □

Example 5.36

Let $f : [0, \infty) \to \mathbb{R}$ be given by

$$f(x) = \frac{(-1)^{n+1}}{n} \quad (n-1 \le x < n)$$

for $n = 1, 2, \dots$. Show that $\int_0^\infty f$ is convergent, but not absolutely convergent.

Solution

Let $k \in (0, \infty)$. Then there exists n in $\mathbb{N}$ such that $n - 1 \le k \le n$. Since f has a constant value in $[n-1, n)$, the integral $I_k = \int_0^k f$ has a value lying between $I_{n-1} = \int_0^{n-1} f$ and $I_n = \int_0^n f$. (If n is even, then $I_n \le I_k \le I_{n-1}$; if n is odd, then $I_{n-1} \le I_k \le I_n$.) Let S be the sum of the alternating harmonic series

$$1 - \frac{1}{2} - \frac{1}{3} + \cdots .$$

Since, for any positive integer n,

$$\int_0^n f = \int_0^1 f + \int_1^2 f + \cdots + \int_{n-1}^n f$$
$$= 1 - \frac{1}{2} + \cdots + (-1)^{n+1} \frac{1}{n},$$

we can, for any given $\epsilon > 0$, choose N large enough so that $|I_n - S| < \epsilon$ for all $n \ge N - 1$ For all $K \ge N$, suppose that $m - 1 \le K \le m$. Then

$$|I_K - S| \le \max \{|I_m - S|, |I_{m-1} - S|\} < \epsilon$$

and we conclude that $\int_0^\infty f$ is convergent.

By contrast,

$$\int_0^N |f| = 1 + \frac{1}{2} + \cdots + \frac{1}{N},$$

and this increases without limit as $N \to \infty$. Hence $\int_0^\infty f$ is not absolutely convergent. $\square$

We close this section with a theorem concerning certain infinite series of positive terms. This is the **integral test**, referred to in connection with Theorem 2.25, and gives an immediate proof of that theorem in the difficult case where $1 < n < 2$.

Theorem 5.37 (The Integral Test)

Let ϕ be a positive, decreasing function with domain $[1, \infty)$. Then the series $\sum_{n=1}^\infty$ is convergent if and only if the integral $\int_1^\infty \phi(x)\, dx$ is convergent.

Proof

Note first that ϕ, being monotonic decreasing, is Riemann integrable by Theorem 5.8 in every finite interval $[1, b]$. Let $N > 1$ be an integer. For all integers k such that $1 \leq k \leq N + 1$, and for all x in $[k, k + 1]$, we have $\phi(k + 1) \leq \phi(x) \leq \phi(k)$, and so

$$\phi(k + 1) \leq \int_k^{k+1} \phi(x)\, dx \leq \phi(k)\,.$$

It follows that

$$\phi(2) + \phi(3) + \cdots + \phi(N) \leq \int_1^N \phi(x)\, dx$$
$$\leq \phi(1) + \phi(2) + \cdots + \phi(N - 1)\,.$$

If $\sum_{n=1}^{\infty} \phi(n)$ is convergent, with sum S (say), then, for every integer $N > 1$,

$$\int_1^N \phi(x)\, dx \leq \phi(1) + \phi(2) + \cdots + \phi(N - 1) \leq S\,,$$

and it follows that $\int_1^{\infty} \phi(x)\, dx$ is convergent. Conversely, if $\sum_{n=1}^{\infty} \phi(n)$ is divergent, then

$$\int_1^N \phi(x)\, dx \geq \phi(2) + \phi(3) + \cdots + \phi(N),$$

and so can be made arbitrarily large by taking N sufficiently large. Thus $\int_1^{\infty} \phi(x)\, dx$ is divergent. □

Remark 5.38

We have concentrated on integrals with upper limit ∞, but it is easy to modify the techniques to cope with integrals with lower limit $-\infty$.

EXERCISES

5.22 Investigate the convergence of

$$\int_0^{\infty} \frac{x\, dx}{\sqrt{x^6 + 1}}, \quad \int_0^{\infty} \frac{(2x + 1)\, dx}{3x^2 + 4\sqrt{x} + 7}\,.$$

5.23 For which values of K (> 0) do the integrals

$$\int_2^{\infty} \left(\frac{Kx}{x^2 + 1} - \frac{1}{2x + 1} \right) dx, \quad \int_0^{\infty} \left(\frac{1}{\sqrt{2x^2 + 1}} - \frac{K}{x + 1} \right) dx$$

converge?

5.24 Give an example of a continuous (not monotonic) function f with the property that $\sum_{n=1}^{\infty} f(n)$ converges and $\int_0^{\infty} f(x)\,dx$ diverges.

5.25 Consider $I = \int_1^{\infty} (\sin x/x)\,dx$.

a) Show that

$$\int_1^K \frac{\sin x}{x}\,dx = \left[\frac{-\cos x}{x}\right]_1^K - \int_1^K \frac{\cos x}{x^2}\,dx \,,$$

and deduce that the integral I is convergent.

b) Show that, for all integers $k \geq 1$

$$\int_{2k\pi}^{(2k+1)\pi} \left|\frac{\sin x}{x}\right|\,dx \geq \frac{2}{(2k+1)\pi} \,, \quad \int_{(2k-1)\pi}^{2k\pi} \left|\frac{\sin x}{x}\right|\,dx \geq \frac{2}{2k\pi} \,,$$

and deduce that I is not absolutely convergent.

5.7 Improper Integrals of the Second Kind

The other way in which an integral $\int_a^b f$ can be improper is if the function f is bounded in every closed interval contained in (a, b) but tends to $\pm\infty$ at a or b or both. This gives us an **improper integral of the second kind**. (If $f(x) \to \pm\infty$ at some point c in (a, b) then we can consider $\int_a^c f$ and $\int_c^b f$ separately and so reduce the problem to the one we have described.)

To be definite, suppose that $f(x) \to \infty$ as $x \to b-$, but that f is bounded and Riemann integrable in the interval $[a, c]$ for every c in (a, b). We say that the integral $\int_a^b f$ is **convergent** if

$$\lim_{c \to b-} \int_a^c f$$

exists. If the limit exists, we define $\int_a^b f$ by

$$\int_a^b f = \lim_{c \to b-} \int_a^c f \,.$$

Similarly, if $f(x) \to \infty$ as $x \to a+$ we define $\int_a^b f$ by

$$\int_a^b f = \lim_{d \to a+} \int_d^b f \,,$$

if this limit exists.

Example 5.39

Investigate the convergence of

$$\int_0^1 \frac{dx}{\sqrt{1-x^2}}.$$

Solution

Here $f(x) \to \infty$ as $x \to 1-$. Now, for every c in $(0,1)$,

$$\int_0^c \frac{dx}{\sqrt{1-x^2}} = \left[\sin^{-1} x\right]_0^c = \sin^{-1} c.$$

Since $\sin^{-1}$ is continuous in $[-1,1]$,

$$\lim_{c \to 1-} \sin^{-1} c = \sin^{-1} 1 = \frac{\pi}{2}.$$

Thus the integral is convergent, and

$$\int_0^1 \frac{dx}{\sqrt{1-x^2}} = \frac{\pi}{2}.$$

□

Theorem 5.40

Let $\alpha > 0$. The improper integral

$$\int_0^1 x^{-\alpha} \, dx$$

is convergent if and only if $\alpha < 1$.

Proof

If $\alpha \neq 1$, then

$$\int_\epsilon^1 x^{-\alpha} \, dx = \left[\frac{x^{-\alpha+1}}{-\alpha+1}\right]_\epsilon^1 = \frac{1}{1-\alpha}\left(1 - \epsilon^{1-\alpha}\right).$$

If $0 < \alpha < 1$ then $1 - \alpha > 0$ and so $\epsilon^{1-\alpha} \to 0$ as $\epsilon \to 0+$. So the integral converges, and

$$\int_0^1 x^{-\alpha} \, dx = \frac{1}{1-\alpha}.$$

If $\alpha > 1$, then $1 - \alpha < 0$ and so $\epsilon^{1-\alpha} \to \infty$ as $\epsilon \to 0+$. So the integral is divergent.

It remains to consider the case $\alpha = 1$. Let k be a positive integer. For every x in the interval $[1/(k+1), 1/k]$ we have

$$k \le \frac{1}{x} \le k+1,$$

and so

$$\int_{1/(k+1)}^{1/k} \frac{dx}{x} \ge k\left(\frac{1}{k} - \frac{1}{k+1}\right) = \frac{1}{k+1}.$$

Thus, for all integers $N > 1$,

$$\int_{1/N}^{1} \frac{dx}{x} = \int_{1/2}^{1} \frac{dx}{x} + \int_{1/3}^{1/2} \frac{dx}{x} + \cdots + \int_{1/(N-1)}^{1/N} \frac{dx}{x}$$

$$\ge \frac{1}{2} + \frac{1}{3} + \cdots + \frac{1}{N-1}.$$

Since the harmonic series is divergent, we can make $\int_{1/N}^{1} (dx/x)$ arbitrarily large by taking N large enough. We conclude that $\int_{0}^{1} (dx/x)$ is divergent. $\quad\square$

It is frequently possible to assert that an improper integral of this kind exists without being able to find its value. Here the asymptotic version of the comparison test is the most useful technique. If $f(x) \to \infty$ and $g(x) \to \infty$ as $x \to a+$, we say that $f(x) \asymp g(x)$ as $x \to a+$ if there exists $K > 0$ such that

$$\lim_{x \to a+} \frac{f(x)}{g(x)} = K,$$

and that $f(x) \sim g(x)$ as $x \to a+$ if

$$\lim_{x \to a+} \frac{f(x)}{g(x)} = 1.$$

We can replace $a+$ by $b-$ in both these definitions.

Theorem 5.41

Let $f, g \in \mathcal{R}[a, c]$ for all c in (a, b), let $f(x), g(x) \to \infty$ as $x \to b-$, and suppose that $f(x) \asymp g(x)$ as $x \to b-$. Then $\int_{a}^{b} f$ is convergent if and only if $\int_{a}^{b} g$ is convergent.

Proof

There exists $K > 0$ such that

$$\lim_{x \to b-} \frac{f(x)}{g(x)} = K.$$

Choosing $\epsilon = K/2$, we can say that there exists $\delta > 0$ such that

$$\left| \frac{f(x)}{g(x)} - K \right| < \frac{K}{2} \tag{5.27}$$

for all x in $(b - \delta, b)$, and we may also suppose that δ is chosen so that $f(x)$ and $g(x)$ are positive in $(b - \delta, b)$. From (5.27) we have, for all x in $(b - \delta, b)$,

$$\frac{K}{2} < \frac{f(x)}{g(x)} < \frac{3K}{2} . \tag{5.28}$$

Suppose first that $\int_a^b g$ converges. From (5.28) we have that $f(x) < (3K/2)g(x)$ for all x in $(b - \delta, b)$. For convenience, denote $\int_{b-\delta/2}^c f$ by $F(c)$. Then, for all c in $[b - \delta/2, b)$,

$$F(c) \le \frac{3K}{2} \int_{b-\delta/2}^c g \le \int_{b-\delta/2}^b g = M \text{ (say)} .$$

Since F is an increasing function in $[b - \delta/2, b)$, and is bounded above by M, we deduce that $\lim_{c \to b-} F(c)$ exists. Thus $\int_a^b f$ is convergent, since

$$\int_a^b f = \int_a^{b-\delta/2} f + \lim_{c \to b-} F(c) .$$

Conversely, if we assume that $\int_a^b f$ is convergent, we use (5.28) to obtain $g(x) < (2/K)f(x)$ $\left(x \in (b - \delta, b) \right)$, and then argue as before. $\square$

Remark 5.42

It is easy to modify our proofs and arguments to deal with the case where $f(x) \to \infty$ as $x \to a+$, or where $f(x) \to -\infty$ either as $x \to a+$ or as $x \to b-$.

Example 5.43

Investigate the convergence of

$$I = \int_0^{\pi/2} \frac{dx}{\sin x} .$$

Solution

We compare $1/\sin x$ with $1/x$. Since

$$\lim_{x \to 0+} \frac{1/x}{1/\sin x} = \lim_{x \to 0+} \frac{\sin x}{x} = 1,$$

we have $1/\sin x \sim 1/x$ as $x \to 0+$. Since $\int_0^{\pi/2} (dx/x)$ is divergent, we deduce that I is divergent also. $\square$

Example 5.44

Investigate the convergence of

$$J = \int_0^1 \frac{dx}{\sqrt{1-x^3}}.$$

Solution

We use the factorization $1 - x^3 = (1 - x)(1 + x + x^2)$, and compare with $1/\sqrt{1-x}$. Since

$$\lim_{x \to 1-} \frac{1/\sqrt{1-x}}{1/\sqrt{1-x^3}} = \lim_{x \to 1-} \frac{\sqrt{1-x^3}}{\sqrt{1-x}}$$

$$= \lim_{x \to 1-} \sqrt{1+x+x^2} = \sqrt{3},$$

we have that

$$\frac{1}{\sqrt{1-x^3}} \asymp \frac{1}{\sqrt{1-x}} \quad \text{as } x \to 1-.$$

Now,

$$\int_0^1 \frac{dx}{\sqrt{1-x}} = \left[-2(1-x)^{1/2}\right]_0^1 = 2,$$

and so we conclude that J is convergent. $\qquad \square$

Example 5.45

Investigate the convergence of

$$I = \int_0^1 \sin\left(\frac{1}{x}\right) dx.$$

Solution

This example is rather different, since $\sin(1/x)$, being oscillatory, has no limit as $x \to 0+$. It is, however, still legitimate to ask whether

$$I(c) = \int_c^1 \sin\left(\frac{1}{x}\right) dx$$

has a limit as $c \to 0+$. In $I(c)$ make the substitution $u = 1/x$ to obtain

$$I(c) = \int_1^{1/c} \frac{\sin u \, du}{u^2}.$$

Since the integral $\int_1^\infty (\sin u/u^2) \, du$ is (absolutely) convergent, the original integral I does converge. $\qquad \square$

EXERCISES

5.26 Investigate the convergence of

$$\int_0^{\pi/2} \frac{dx}{\sqrt{\sin x}}, \quad \int_0^{\pi/2} \frac{\sin x \, dx}{x^2}, \quad \int_0^{\pi/2} \frac{dx}{x - \sin x}.$$

5.27 Show that

$$\int_0^\infty \frac{\sin x}{x} \, dx = \int_0^\infty \frac{\sin^2 x}{x^2} \, dx.$$

6

The Logarithmic and Exponential Functions

6.1 A Function Defined by an Integral

It is convenient at this stage to introduce two of the most important functions in mathematics. In a huge range of applications, from the discharge of a capacitor to the population growth of bacteria, the exponential function plays a crucial role, and "log–log" graphs are a crucial part of the methodology over a wide area of experimental science.

But first let us remind ourselves of what we mean by a logarithm. If $a^x = b$ we say that $x = \log_a b$, the **logarithm of b to the base** a. In words, $\log_a b$ is the power to which a must be raised to obtain b. There are difficulties about this definition. For example, we know what a^x means only if x is rational; and – a related difficulty – it is not clear that $\log_a b$ is defined for every b. However, these difficulties will vanish very soon now.

We begin with the simple observation that the formula

$$\int x^n \, dx = \frac{x^{n+1}}{n+1}$$

is not valid when $n = -1$. On the other hand, the function $x \mapsto 1/x$ is continuous in any interval $[a, b]$ not containing 0, and so (by Theorem 5.9) is Riemann integrable in any such interval. Let us therefore define a new function L by

$$L(x) = \int_1^x \frac{dt}{t} \quad (x > 0).$$

Certain properties of L are immediate. By the fundamental theorem (Theorem 5.18), L is differentiable (and so certainly continuous) and

$$L'(x) = \frac{1}{x}. \qquad (6.1)$$

Thus L is an increasing function for all x in $(0, \infty)$. Notice also that

$$L(1) = 0.$$

The crucial property of the function L is given in the following theorem:

Theorem 6.1

With the above definitions, for all x, y in $(0, \infty)$,

$$L(xy) = L(x) + L(y), \quad L(1/x) = -L(x).$$

Proof

By Theorem 5.12 and Remark 5.13,

$$L(xy) = \int_1^x \frac{dt}{t} + \int_x^{xy} \frac{dt}{t} = L(x) + J \text{ (say)}.$$

Now, in the integral J, let $u = t/x$. Then $u = 1$ when $t = x$, and $u = y$ when $t = xy$. Also $du = (1/x)dt$. Hence

$$J = \int_1^y \frac{x\,du}{xu} = \int_1^y \frac{du}{u} = L(y),$$

and so $L(xy) = L(x) + L(y)$, as required.

The second statement follows immediately, since

$$L(x) + L(1/x) = L(1) = 0.$$

$\square$

Consider the dissection $D = \{1, \frac{3}{2}, 2\}$ of the interval $[1, 2]$. Since the reciprocal function $R : x \mapsto 1/x$ is continuous and monotonic decreasing,

$$\sup_{(1,3/2)} R = 1, \quad \sup_{(3/2,2)} R = R(3/2) = 2/3.$$

Hence

$$L(2) \le \mathcal{U}(R, D) = \frac{1}{2}\left(1 + \frac{2}{3}\right) = \frac{5}{6} < 1.$$

Next, consider the dissection $D' = \{1, \frac{5}{4}, \frac{3}{2}, \frac{7}{4}, 2, \frac{9}{4}, \frac{5}{2}, \frac{11}{4}, 3\}$ of the interval $[1,3]$. Here

$$\inf_{(1,5/4)} R = 4/5, \quad \inf_{(5/4,3/2)} R = 2/3, \dots, \quad \inf_{(11/4,3)} R = 1/3,$$

and an easy (though tedious) calculation shows that

$$L(3) \geq \mathcal{L}(R, D') = \frac{1}{4}\left(\frac{4}{5} + \frac{2}{3} + \frac{4}{7} + \frac{1}{2} + \frac{4}{9} + \frac{2}{5} + \frac{4}{11} + \frac{1}{3}\right) = \frac{28,271}{27,720} > 1.$$

It follows that there exists a unique number[1] e, lying between 2 and 3, such that

$$L(e) = \int_1^e \frac{dt}{t} = 1.$$

We have seen that $\int_1^\infty (dt/t)$ diverges, and so

$$L(x) \to \infty \quad \text{as} \quad x \to \infty.$$

Since $1/x \to \infty$ as $x \to 0+$,

$$L(x) = -L(1/x) \to -\infty \quad \text{as} \quad x \to 0+ .$$

The graph of L is given in Fig. 6.1.

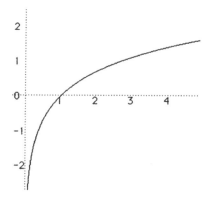

Figure 6.1. The graph of $\log x$

---

[1] We shall see eventually that e is the same number as that encountered in Example 2.13.

EXERCISES

6.1 Show that, for all n in $\mathbb{N}$,

$$L(2^n) > \frac{n}{2}.$$

6.2 The Inverse Function

Since L is differentiable and increasing throughout its domain, there is an increasing differentiable inverse function $L^{-1} : \mathbb{R} \to (0, \infty)$, which for the moment we shall denote by E. Thus

$$E(L(x)) = x \ (x \in (0, \infty)), \quad L(E(x)) = x \ (x \in \mathbb{R}).$$

The function E has the friendly property of being its own derivative: by Theorem 4.15 and (6.1),

$$E'(x) = (L^{-1})'(x) = \frac{1}{L'(L^{-1}(x))} = L^{-1}(x) = E(x).$$

The graph of E is given in Fig. 6.2.

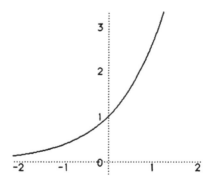

Figure 6.2. The graph of $E(x)$

Since $L(1) = 0$ and $L(e) = 1$, we have

$$E(0) = 1, \quad E(1) = e.$$

From Theorem 6.1 we can deduce the crucial property of the function E:

Theorem 6.2

For all x, y in $\mathbb{R}$,

$$E(x+y) = E(x)E(y), \quad E(-x) = \frac{1}{E(x)} \,.$$

Proof

For all x, y in $\mathbb{R}$,

$$\begin{aligned}
E(x+y) &= E\big[L\big(E(x)\big) + L\big(E(y)\big)\big] \\
&= E\big[L\big(E(x)E(y)\big)\big] \text{ (by Theorem 6.1)} \\
&= E(x)E(y) \,.
\end{aligned}$$

The second statement follows immediately, since

$$E(x)E(-x) = E\big(x + (-x)\big) = E(0) = 1 \,.$$

$\square$

From the theorem we deduce in particular that, for all x in $\mathbb{R}$,

$$[E(x)]^2 = E(x)E(x) = E(x + x) = E(2x) \,,$$

and, more generally, that

$$[E(x)]^n = E(nx) \quad (n \in \mathbb{N}) \,. \tag{6.2}$$

In particular, putting $x = 1$, we have

$$E(n) = [E(1)]^n = e^n \quad (n \in \mathbb{N}) \,. \tag{6.3}$$

This holds in fact for *all* integers: clearly $E(0) = 1 = e^0$, and, if $n = -m$, with $m \in \mathbb{N}$, then

$$E(n) = E(-m) = \frac{1}{E(m)} = \frac{1}{e^m} = e^{-m} = e^n \,.$$

The formula even extends to all rational numbers. Let $q = m/n$, where $m \in \mathbb{Z}$ and $n \in \mathbb{N}$; then

$$\begin{aligned}
[E(q)]^n &= E(nq) \quad \text{(by (6.2))} \\
&= E(m) = e^m \,,
\end{aligned}$$

and so

$$E(q) = \sqrt[n]{e^m} = e^{m/n} = e^q \,.$$

So far we have no meaning for a^x when a is a positive real number and x is irrational. We shall come to the case of a general a shortly, but, given that $e^q = E(q)$ for every rational, it seems reasonable to *define* e^x as $E(x)$ for every real number. At this point we shall drop the temporary notation $E(x)$ and write either e^x or $\exp x$. The function is always called the **exponential function**.

If $y = e^x$ then $x = L(y)$. We therefore drop the temporary notation $L(y)$ and instead[2] write $\log_e y$, or just $\log y$. It is sometimes called the **natural logarithm**.

Remark 6.3

Before the advent of electronic calculators, logarithms (to the base 10) were used as an aid to calculation, and students carried four-figure "log tables" wherever they went. To calculate (say) 325.7×48.43 one looked up the two logarithms (approximately 2.5128 and 1.6851 respectively), added the two logarithms, obtaining 4.1979, and then found $10^{4.1979}$ (which, to four significant figures, is 15,770) by looking up a table of "antilogarithms". The answer is not quite accurate (the correct answer is 15,773.651) but for most practical purposes an error of 0.02% is not significant. The use of logarithms as an aid to calculation has a long history, going back to Napier[3] and Briggs[4] in the sixteenth century. Modern technology has turned the technique into a mere curiosity, but the *idea* of a logarithm remains as important as ever.

In the new notation,

$$e^{\log x} = x \ (x > 0), \quad \log(e^x) = x \ (x \in \mathbb{R}),$$

$$\int_1^x \frac{dt}{t} = \log x, \quad \frac{d}{dx}(\log x) = \frac{1}{x}, \quad \frac{d}{dx}(e^x) = e^x, \quad \int e^x \, dx = e^x.$$

If a is a positive real number, we now *define*

$$a^x = e^{x \log a}, \quad \log_a(x) = \frac{\log x}{\log a}.$$

The differentiation formula for x^n survives this last extension:

Example 6.4

Let $f(x) = x^u$, where u is any real number. Show that $f'(x) = ux^{u-1}$.

---

[2] The notation $\ln y$ is also used.
[3] John Napier, 1550–1617.
[4] Henry Briggs, 1561–1631.

Solution

By definition, $f(x) = e^{u \log x}$, and so

$$f'(x) = e^{u \log x} \cdot \frac{u}{x} = x^u \cdot \frac{u}{x} = u x^{u-1} \,.$$

□

The next example shows in particular that the number e defined above as $E(1)$ is the same as that encountered a long time ago in Example 2.13.

Example 6.5

Show that, for all x in $\mathbb{R}$,

$$\lim_{n \to \infty} \left(1 + \frac{x}{n}\right)^n = e^x \,.$$

Solution

If $x = 0$ the result is immediate. Now suppose that $x \neq 0$, and define a function f by

$$f(t) = \log(1 + xt) \,.$$

The domain of this function is $\{t : 1 + xt > 0\}$, which equals $(-1/x, \infty)$ if x is positive, and $(-\infty, -1/x)$ is x is negative. In either case the domain includes 0. Since

$$f'(t) = \frac{x}{1 + xt}$$

it follows that $f'(0) = x$. That is,

$$\lim_{h \to 0} \frac{\log(1 + xh)}{h} = x \,.$$

We now replace h by $1/n$, where n is a positive integer, and deduce that

$$x = \lim_{n \to \infty} n \log \left(1 + \frac{x}{n}\right) = \lim_{n \to \infty} \log \left[\left(1 + \frac{x}{n}\right)^n\right] \,.$$

Since the function $x \mapsto e^x$ is continuous, we can apply it to both sides, and obtain the required result, that

$$\lim_{n \to \infty} \left(1 + \frac{x}{n}\right)^n = e^x \,.$$

□

The final example for this section explores the strong connection between the logarithmic function and the harmonic series.

Example 6.6

Let

$$H_n = 1 + \frac{1}{2} + \cdots + \frac{1}{n},$$

and define $\gamma_n = H_n - \log n$. Show that the sequence (γ_n) tends to a limit[5] γ, where $0 < \gamma \leq 1$. Deduce that

$$1 - \frac{1}{2} + \frac{1}{3} - \cdots = \log 2.$$

Solution

In each interval $[k, k+1]$ (where $k \in \mathbb{N}$) we have

$$\frac{1}{k+1} \leq \frac{1}{x} \leq \frac{1}{k}.$$

Hence

$$\frac{1}{k+1} \leq \int_k^{k+1} \frac{dx}{x} \leq \frac{1}{k}. \qquad (6.4)$$

Summing from $k = 1$ to $n-1$ gives

$$\frac{1}{2} + \cdots + \frac{1}{n} \leq \int_1^n \frac{dx}{x} \leq 1 + \frac{1}{2} + \cdots + \frac{1}{n-1}$$

which can be rewritten as

$$H_n - 1 \leq \log n \leq H_n - \frac{1}{n}.$$

Hence

$$-1 \leq \log n - H_n \leq -\frac{1}{n},$$

and so

$$\frac{1}{n} \leq \gamma_n \leq 1.$$

Thus the sequence (γ_n) is a sequence of positive numbers, bounded above by 1.

We show that (γ_n) is a monotonic decreasing sequence:

$$\gamma_{n+1} - \gamma_n = (H_{n+1} - H_n) - (\log(n+1) - \log n)$$
$$= \frac{1}{n+1} - \int_n^{n+1} \frac{dx}{x}$$
$$\leq 0 \quad \text{(by (6.4))}.$$

---

[5] The number γ is known as **Euler's constant**, after Leonard Euler (1707–1783). Its value is approximately 0.5772, and it is still unknown whether it is rational or irrational.

It then follows by Theorem 2.9 that (γ_n) has a limit γ, where $0 < \gamma \le 1$.

Look now at the sum to $2n$ terms of the alternating harmonic series:

$$
\begin{aligned}
S_{2n} &= 1 - \frac{1}{2} + \frac{1}{3} - \cdots - \frac{1}{2n} \\
&= \left(1 + \frac{1}{2} + \cdots + \frac{1}{2n}\right) - 2\left(\frac{1}{2} + \frac{1}{4} + \cdots + \frac{1}{2n}\right) \\
&= H_{2n} - \left(1 + \frac{1}{2} + \cdots + \frac{1}{n}\right) \\
&= H_{2n} - H_n = (\log(2n) + \gamma_{2n}) - (\log n + \gamma_n) \\
&= \log 2 + \gamma_{2n} - \gamma_n \\
&\to \log 2 + \gamma - \gamma = \log 2 \quad \text{as } n \to \infty.
\end{aligned}
$$

Since

$$
S_{2n+1} = S_{2n} - \frac{1}{2n+1} \to \log 2 + 0
$$

as $n \to \infty$, we deduce that the sum of the series is $\log 2$. $\qquad\square$

Remark 6.7

In Chapter 2 (see the argument following eq. (2.5)), when assessing the rate of divergence of the harmonic series, we remarked that a more accurate estimate could be found. We now know that $H_n \sim \log n$, and so $H_n > 100$ when $n > e^{100} \approx 2.7 \times 10^{43}$.

EXERCISES

6.2 Show that, for all $u > 0$,

$$
1 - u < \frac{1}{1+u} < 1,
$$

and deduce that

$$
x - \frac{1}{2}x^2 < \log(1+x) < x \quad (x > 0).
$$

Hence, or otherwise, deduce that

$$
\lim_{x \to 0} \frac{\log(1+x)}{x} = 1.
$$

6.3 Let
$$f(x) = x - 1 - \log x, \quad g(x) = \log x - 1 + \frac{1}{x},$$
where $x > 0$. By considering the signs of $f'(x)$ and $g'(x)$ for $x < 1$ and for $x > 1$, show that, for all x in $(0, \infty) \setminus \{1\}$,
$$1 - \frac{1}{x} < \log x < x - 1.$$

6.4 Using Taylor's Theorem, show that, for all $x > 0$,
$$x - \frac{1}{2}x^2 < \log(1 + x) < x - \frac{1}{2}x^2 + \frac{1}{3}x^3.$$

6.5 Let
$$L(m, n) = \int x^m (\log x)^n \, dx \quad (m, n \geq 0).$$
Show that, for all $m \geq 0$ and $n \geq 1$,
$$L(m, n) = \frac{x^{m+1}(\log x)^n}{m + 1} - \frac{n}{m + 1} L(m, n - 1),$$
and hence determine $\int x (\log x)^3$.

6.6 Determine
$$\lim_{x \to 0} \frac{\log(\cos ax)}{\log(\cos bx)}.$$

6.7 Show that $e^{-x} > 1 - x$ for all $x \neq 0$ in $\mathbb{R}$, and deduce that
$$1 + x < e^x < \frac{1}{1 - x} \quad (0 < x < 1).$$

6.8 The **hyperbolic functions** cosh and sinh are defined by
$$\cosh x = \frac{1}{2}(e^x + e^{-x}), \quad \sinh x = \frac{1}{2}(e^x - e^{-x}).$$

a) Show that
$$\text{im cosh} = [1, \infty), \quad \text{im sinh} = \mathbb{R}$$
$$\cosh(-x) = \cosh x, \quad \sinh(-x) = -\sinh x.$$
Sketch the graphs.

b) Show that $(\cosh)'x = \sinh x$, $(\sinh)'x = \cosh x$, and that
$$\cosh^2 x - \sinh^2 x = 1,$$
$$\cosh(x + y) = \cosh x \cosh y + \sinh x \sinh y,$$
$$\sinh(x + y) = \sinh x \cosh y + \cosh x \sinh y.$$

c) Show that sinh has an inverse function $\sinh^{-1} : \mathbb{R} \to \mathbb{R}$, and that

$$(\sinh^{-1})'(x) = \frac{1}{\sqrt{1+x^2}} \,.$$

Show also that, if we restrict the domain of cosh to $[0, \infty)$, then there is an inverse function $\cosh^{-1} : [1, \infty) \to [0, \infty)$, and that

$$(\cosh^{-1})'(x) = \frac{1}{\sqrt{x^2-1}} \,.$$

d) Show that

$$\sinh^{-1} x = \log(x + \sqrt{1+x^2}) \quad (x \in \mathbb{R}),$$
$$\cosh^{-1} x = \log(x + \sqrt{x^2-1}) \quad (x \in [1, \infty)).$$

6.9 Determine $\int (1/x \log x)\, dx$, and deduce that the integral

$$\int_2^\infty \frac{dx}{x \log x}$$

is divergent. For $n \geq 3$ let

$$K_n = \frac{1}{3 \log 3} + \cdots + \frac{1}{n \log n} \,.$$

Using the method of Example 6.6, show that

$$\frac{1}{n \log n} \leq K_n - [\log \log n - \log \log 3] \leq \frac{1}{3 \log 3} \,.$$

Show that the sequence (δ_n), where $\delta_n = K_n - \log \log n$, has a limit δ.

How many terms of the divergent series $\sum_{n=2}^\infty (1/n \log n)$ are required to make the sum exceed 5?

6.10 Find a divergent series for which the sum S_n to n terms has the property that $S_n \asymp \log \log \log n$.

6.3 Further Properties of the Exponential and Logarithmic Functions

If we apply the Taylor–Maclaurin Theorem (4.13) to the exponential function, we obtain

$$e^x = e^0 + xe^0 + \frac{x^2}{2!}e^0 + \cdots + \frac{x^{n-1}}{(n-1)!}e^0 + R_n$$

$$= 1 + x + \frac{x^2}{2!} + \cdots + \frac{x^{n-1}}{(n-1)!} + R_n \,,$$

where

$$R_n = \frac{x^n}{n!}e^{\theta x} \quad (\theta \in (0,1))\,.$$

We noted in Example 2.35 that $\lim_{n\to\infty}(x^n/n!) = 0$ for all x. Thus $R_n \to 0$ as $n \to \infty$, and so we have the Taylor–Maclaurin series

$$e^x = e^0 + xe^0 + \frac{x^2}{2!}e^0 + \frac{x^3}{3!}e^0 + \cdots = \sum_{n=0}^{\infty} \frac{x^n}{n!} \,, \tag{6.5}$$

convergent for all values of x.

Remark 6.8

·This rapidly convergent series gives a practical way of computing the number e. We have

$$e = 1 + 1 + \frac{1}{2!} + \frac{1}{3!} + \cdots \,,$$

and after just 12 terms the error is

$$\frac{1}{13!} + \frac{1}{14!} + \cdots < \frac{1}{13!}\left(1 + \frac{1}{14} + \frac{1}{14^2} + \cdots\right)$$

$$= \frac{1}{13!} \cdot \frac{1}{1 - (1/14)} = \frac{14}{13.13!}$$

$$< 2 \times 10^{-10} = 0.0000000002\,.$$

Thus we obtain an estimate 2.718281828 for e, correct to 9 decimal places. The appearance of a repeating decimal is illusory, since, as we shall see in Example 6.10 below, e is irrational.

From the series expansion (6.5) for e^x we can deduce some important properties of the exponential function. We already know that $e^x \to \infty$ as $x \to \infty$, but we can now show that it tends to infinity **faster than any power of** x To be precise, let $x > 1$, and consider a positive power x^α. Let m be any integer such that $m > \alpha$. Then all terms of the series (6.5) are positive; hence

$$e^x > \frac{x^m}{m!}\,,$$

and so

$$\frac{e^x}{x^\alpha} > \frac{x^{m-\alpha}}{m!}\,.$$

Since $x^{m-\alpha}/m! \to \infty$ as $x \to \infty$, we deduce that, for all $\alpha > 0$,

$$x^{-\alpha}e^x \to \infty \quad\text{as } x \to \infty \tag{6.6}$$

or (equivalently)

$$\lim_{x\to\infty} x^\alpha e^{-x} = 0\,. \tag{6.7}$$

More generally, we can easily show that, for all positive α and all positive k,

$$\lim_{x\to\infty} x^\alpha e^{-kx} = 0\,. \tag{6.8}$$

By contrast, we find that $\log x \to \infty$ as $x \to \infty$ more slowly than any positive power of x. Let x be positive, and let x^α be a positive power. Then, putting $y = \log x$ (so that $x = e^y$), and observing that $y \to \infty$ as $x \to \infty$, we have

$$\lim_{x\to\infty} x^{-\alpha}\log x = \lim_{y\to\infty} ye^{-\alpha y} = 0\,. \tag{6.9}$$

From this we can deduce

$$\lim_{x\to 0+} x^\alpha \log x = 0 \tag{6.10}$$

for all positive α; for, putting $y = 1/x$, we see that

$$\lim_{x\to 0+} x^\alpha \log x = \lim_{y\to\infty} \left(-y^{-\alpha}\log y\right) = 0\,.$$

Remark 6.9

The property that the derivative of e^{kx} is ke^{kx} is the key to the importance of the function in applications. For example, in the presence of an adequate food supply and adequate space for expansion, the rate of growth of a population of bacteria is proportional to the number of bacteria, and so is a function $P(t)$ of time, with the property that $P'(t) = kP(t)$. It follows that

$$\frac{d}{dt}\log\big(P(t)\big) = \frac{P'(t)}{P(t)} = k\,,$$

from which it follows that $\log\big(P(t)\big) = kt + C$. Thus

$$P(t) = e^{kt+C} = Ae^{kt}\,,$$

where A is the initial number of bacteria. If we measure t in hours, and make the simplest possible assumption that $A = k = 1$, then we reach the frightening conclusion that after 24 hours the population is approximately 2.65×10^{10}.

Example 6.10

Show that e is irrational.

Solution

Suppose, for a contradiction, that $e = p/q$, where $p, q \in \mathbb{N}$, and let n be an integer such that $n > \max\{q, 3\}$. Then $n!e$ is an integer. From Taylor's Theorem,

$$e = 2 + \frac{1}{2!} + \cdots + \frac{1}{n!} + \frac{1}{(n+1)!}e^{\theta}$$

where $0 < \theta < 1$. Hence

$$\frac{e^{\theta}}{n+1} = n!e - 2n! - \frac{n!}{2!} - \cdots - \frac{n!}{n!}$$

is an integer. On the other hand, $1 < e^{\theta} < e < 3$, and so

$$0 < \frac{e^{\theta}}{n+1} < \frac{3}{n+1} < 1.$$

From this contradiction we deduce that e is irrational.   □

EXERCISES

6.11 Let

$$\Gamma(\alpha) = \int_0^{\infty} e^{-x}x^{\alpha-1}\,dx\,.$$

Show that the integral converges if and only if $\alpha > 0$. Prove also that $\Gamma(\alpha) = (\alpha - 1)\Gamma(\alpha - 1)$ for all $\alpha > 1$, and deduce that $\Gamma(n) = (n-1)!$ for all n in $\mathbb{N}$.

6.12 Show that, for all $\alpha, y > 0$

$$\lim_{x \to \infty} x^{-k}(\log x)^{\alpha} = 0, \quad \lim_{x \to 0+} x^{k}(\log x)^{\alpha} = 0\,.$$

6.13 Show that $\lim_{n \to \infty} n^{1/n} = 1$.

6.14 Show that $\log(x \log x) \sim \log x$ as $x \to \infty$.

6.15 Determine

$$\lim_{x \to 0} \frac{a^x - b^x}{c^x - d^x},$$

where $a, b, c, d > 0$ and $c \neq d$.

6.16 Show that, if $s > 0$ and $n \in \mathbb{N}$,

$$\int_0^\infty e^{-st} t^n \, dt = \frac{n!}{s^{n+1}}, \quad \int_0^\infty e^{-st} \cos at \, dt = \frac{s}{s^2 + a^2},$$

$$\int_0^\infty e^{-st} \sin at \, dt = \frac{a}{s^2 + a^2}.$$

[These are examples of **Laplace**[6] **transforms**, which are the basis of an important technique for solving differential equations. See [1].]

6.17 Let $f(x) = e^{-1/x^2}$ $(x \neq 0)$, $f(0) = 0$. Prove, by induction on n, that $f^{(n)}(x) = P_n(1/x)e^{-1/x^2}$ for all $x \neq 0$, where P_n is a polynomial function. Hence prove that $f^n(0) = 0$ for every n. Deduce that f, despite having derivatives of every order, does not have a convergent Taylor–Maclaurin series, and so is not analytic.

6.18 Prove that $n! > (n/e)^n$.

6.19 By considering the dissection $\{1, 2, \ldots, n\}$ of the interval $[1, n]$ and the corresponding lower and upper sums of the function $x \mapsto \log x$, show that

$$(n-1)! \leq \frac{n^n}{e^{n-1}} \leq n!.$$

---

[6] Pierre Simon Laplace, 1749–1827

7
Sequences and Series of Functions

7.1 Uniform Convergence

Let f be a function whose domain is the closed interval $[a, b]$. We define $\|f\|$, the **norm** of f, by

$$\|f\| = \sup_{[a,b]} |f| \ (= \sup \{|f(x)| : x \in [a, b]\}).\tag{7.1}$$

The norm has the following properties:

Theorem 7.1

Let f, g be functions with domain $[a, b]$, and let $k \in \mathbb{R}$ be a constant.

(i) $\|f\| \geq 0$, and $\|f\| = 0$ if and only if $f = 0$;

(ii) $\|f + g\| \leq \|f\| + \|g\|$;

(iii) $\|kf\| = |k| \, \|f\|$.

Proof

(i) The first part is clear from the definition. Also, if $\|f\| = 0$, then, for all x in $[a, b]$,

$$0 \leq |f(x)| \leq \|f\| = 0,$$

and so $f(x) = 0$ for all x in $[a, b]$ (which of course is what we mean when we write $f = 0$).

(ii) For all x in $[a, b]$,

$$|f + g|(x) = |f(x) + g(x)| \le |f(x)| + |g(x)| \le \sup_{[a,b]} |f| + \sup_{[a,b]} |g| = \|f\| + \|g\|,$$

and from this it follows that

$$\|f + g\| = \sup_{[a,b]} |f + g| \le \|f\| + \|g\|.$$

(iii) Since $|kf| = |k|\,|f|$, it follows that

$$\|kf\| = \sup_{[a,b]} |kf| = \sup_{[a,b]} |k|\,|f| = |k| \sup_{[a,b]} |f| = |k|\,\|f\|.$$

$\square$

We can use this idea to provide a useful notion of the distance between two functions: if f, g are functions with domain $[a, b]$, then the **distance between f and g** is defined as

$$\|f - g\| = \sup_{[a,b]} |f - g| \ (= \sup \{|f(x) - g(x)| : x \in [a, b]\}). \qquad (7.2)$$

This distance function has the following properties:

Theorem 7.2

Let f, g, h be functions with domain $[a, b]$.

(i) $\|f - g\| \ge 0$, and $\|f - g\| = 0$ if and only if $f = g$;

(ii) $\|g - f\| = \|f - g\|$;

(iii) $\|f - h\| \le \|f - g\| + \|g - h\|$.

Proof

Parts (i) and (ii) are immediate. Part (iii) follows easily from Theorem 7.1(ii), since

$$\|f - h\| = \|(f - g) + (g - h)\| \le \|f - g\| + \|f - h\|.$$

This is usually called the **triangle inequality**. If we think of f, g and h as "points" forming the vertices of a triangle, it tells us that the length of one side cannot exceed the sum of the lengths of the other two sides. $\square$

We now apply this idea to obtain a definition of convergence of a sequence of functions. Let (f_n) be a sequence of functions with fixed domain $[a, b]$. Then the sequence (f_n) is said to have **uniform limit** f **in the interval** $[a, b]$, or to **converge uniformly to** f **in the interval** $[a, b]$, if for every $\epsilon > 0$ there exists a positive integer N with the property that $\|f_n - f\| < \epsilon$ for all $n > N$.

Remark 7.3

It is easy to see that, if $(f_n) \to f$ uniformly in $[a, b]$, then it also converges uniformly to f in any interval $[c, d]$ contained in $[a, b]$. This follows from the simple observation that $\sup_{[c,d]} |f_n - f| \leq \sup_{[a,b]} |f_n - f|$.

We are using the word "uniform(ly)" here as a qualifier because this is not the only possible definition. The other kind of convergence is called "pointwise": we say that (f_n) **converges pointwise to** f **in the interval** $[a, b]$ if, for all x in $[a, b]$, the sequence $(f_n(x))$ converges to $f(x)$.

The appropriateness of the word "pointwise" is clear; the appropriateness of the word "uniformly" will be clear shortly. First, however, we shall satisfy ourselves that the two notions are distinct, by showing that uniform convergence is a stronger property than pointwise convergence.

Theorem 7.4

Let (f_n) be a sequence of functions with domain $[a, b]$. If (f_n) converges uniformly to f, then (f_n) converges pointwise to f.

Proof

For all $\epsilon > 0$ there exists a positive integer N such that $\|f_n - f\| < \epsilon$ for all $n > N$. Let $x \in [a, b]$. Then

$$|f_n(x) - f(x)| \leq \sup_{[a,b]} |f_n(x) - f(x)| = \|f_n - f\| < \epsilon$$

for all $n > N$. Hence $(f_n(x)) \to f(x)$. $\square$

The converse of this theorem is not true, as the following example demonstrates. Let $f_n(x) = x^n$ $(x \in [0, 1])$. If $x \in [0, 1)$, then $f_n(x) \to 0$ as $n \to \infty$; if $x = 1$, then $x^n = 1$ for all n, and so $(f_n(x)) \to 1$. Thus (f_n) converges pointwise to the function f defined by

$$f(x) = \begin{cases} 0 & \text{if } 0 \leq x < 1 \\ 1 & \text{if } x = 1. \end{cases}$$

The convergence is not uniform, however. For each n the function f_n is continuous, and increases from 0 to 1.

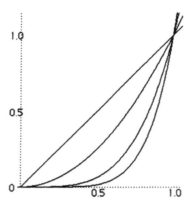

Figure 7.1. The functions f_n

So there exists a unique number p_n (namely $p_n = 1/\sqrt[n]{2}$) such that $f_n(p_n) = 1/2$. It follows that, for all $n \geq 1$,

$$\|f_n = f\| = \sup_{[0,1]} |f_n(x) - f(x)| \geq |f_n(p_n) - f(p_n)| = \frac{1}{2},$$

and so (f_n) definitely does *not* converge uniformly to f. In fact (f_n) does not have a uniform limit at all, for if it did, then Theorem 7.4 would force the limit to be f, and we have seen that this is not so.

It is worth exploring this example a little further to find the reason for the distinction between pointwise and uniform convergence. Let $p \in (0, 1)$, and let $0 < \epsilon < 1$. Then

$$|f_n(p) - f(p)| = p^n < \epsilon$$

if and only if $n > \lceil \log \epsilon / \log p \rceil$, the smallest integer N such that $N \geq \log \epsilon / \log p$. (Notice that both $\log \epsilon$ and $\log p$ are negative.) To be definite, suppose that $\epsilon = 0.001$, so that $\log \epsilon \approx -6.9$. Then we have the following table:

p	0.9	0.99	0.999	0.9999
N	66	688	6,905	69,075

As p approaches 1 we need ever larger values of n to ensure that $|f_n(p) - f(p)| < 0.001$ for all $n > N$. There is no N that will work "uniformly" for every p in $[0, 1)$.

We remark that, for a fixed p in $(0, 1)$, the sequence (f_n) *does* converge uniformly to f (the zero function) in every closed interval $[0, p]$. To see this, for a given ϵ choose $N = \lceil \log \epsilon / \log p \rceil$. Then, for all $n > N$

$$\|f_n - f\| = \sup_{[0,p]} |f_n| = \sup_{[0,p]} x^n = p^n < \epsilon,$$

which is exactly what we require. The point here is that there is a definite "worst case", namely p itself. If we consider the whole interval $[0,1]$ then we have no worst case, only worse and worse cases.

Another way of seeing the difference between pointwise and uniform convergence, which you may or may not find helpful, is to use quantifier symbols $\forall$ (for all) and $\exists$ (there exists). The sequence (f_n) has pointwise limit f in $[a,b]$ if

$$(\forall \epsilon > 0)(\forall x \in [a,b])(\exists N \in \mathbb{N})(\forall n > N)[|f_n(x) - f(x)| < \epsilon].$$

It has uniform limit f in $[a,b]$ if

$$(\forall \epsilon > 0)(\exists N \in \mathbb{N})(\forall x \in [a,b])(\forall n > N)[|f_n(x) - f(x)| < \epsilon].$$

The change in the order of the quantifiers is crucial. For uniform convergence we must find a single N that works for every x in $[a,b]$. For pointwise convergence the N may vary for different values of x, a point well illustrated in the example above.

The fact that in the example above the convergence is not uniform could have been deduced indirectly from the next theorem, for in that example each f_n is continuous on $[0,1]$, but f is not.

Theorem 7.5

Let (f_n) be a sequence of functions with common domain $[a,b]$, and suppose that $(f_n) \to f$ uniformly on $[a,b]$. If each f_n is continuous on $[a,b]$, then so is f.

Proof

Let $c \in [a,b]$, and let $\epsilon > 0$ be given. There exists a positive integer N such that $\|f_n - f\| < \epsilon/3$ for all $n > N$. Now choose and fix an integer $M > N$. ($M = N+1$ will do very well.) Then, in particular, $\|f_M - f\| < \epsilon/3$. The function f_M is continuous, and so there exists $\delta > 0$ such that $|f_M(x) - f_M(c)| < \epsilon/3$ for all x in $[a,b] \cap (c - \delta, c + \delta)$. Hence, for all x in $[a,b] \cap (c - \delta, c + \delta)$,

$$\begin{aligned}
|f(x) - f(c)| &= \left| (f(x) - f_M(x)) + (f_M(x) - f_M(c)) + (f_M(c) - f(c)) \right| \\
&\le |f(x) - f_M(x)| + |f_M(x) - f_M(c)| + |f_M(c) - f(c)| \\
&\le \|f_M - f\| + |f_M(x) - f_M(c)| + \|f_M - f\| \\
&< \frac{\epsilon}{3} + \frac{\epsilon}{3} + \frac{\epsilon}{3} = \epsilon.
\end{aligned}$$

Hence f is continuous at each c in $[a, b]$, as required.   □

Next, we have a theorem about the integration of a uniformly convergent sequence.

Theorem 7.6

Let (f_n) be a sequence of functions in $\mathcal{R}[a, b]$, and suppose that $(f_n) \to f$ uniformly on $[a, b]$. Then $f \in \mathcal{R}[a, b]$ and

$$\lim_{n \to \infty} \int_a^b f_n = \int_a^b f .$$

Proof

We show first that $f \in \mathcal{R}[a, b]$. Since $f_n \to f$ uniformly in $[a, b]$, there exists N such that $\|f_n - f\| < \epsilon/3(b-a)$ for all $n > N$. Let $D = \{a = x_0, x_1, \ldots, x_m = b\}$ be a dissection of $[a, b]$. For all $n > N$, we have that

$$| \sup_{(x_{i-1}, x_i)} f_n - \sup_{(x_{i-1}, x_i)} f| \leq \frac{\epsilon}{3(b-a)} ,$$

and so

$$|\mathcal{U}(f_n, D) - \mathcal{U}(f, D)| = \left| \sum_{i=1}^m (\sup_{(x_{i-1}, x_i)} f_n - \sup_{(x_{i-1}, x_i)} f)(x_i - x_{i-1}) \right|$$

$$\leq \sum_{i=1}^m | \sup_{(x_{i-1}, x_i)} f_n - \sup_{(x_{i-1}, x_i)} f|(x_i - x_{i-1})$$

$$\leq \frac{\epsilon}{3} .$$

A similar argument shows that

$$|\mathcal{L}(f_n, D) - \mathcal{L}(f, D)| \leq \frac{\epsilon}{3} .$$

For any given n we may choose D so that

$$\mathcal{U}(f_n, D) - \mathcal{L}(f_n, D) \leq \frac{\epsilon}{3} ,$$

and it then follows that

$$\mathcal{U}(f, D) - \mathcal{L}(f, D)$$
$$= [\mathcal{U}(f, D) - \mathcal{U}(f_n, D)] + [\mathcal{U}(f_n, D) - \mathcal{L}(f_n, D)] + [\mathcal{L}(f_n, D) - \mathcal{L}(f, D)]$$
$$\leq |\mathcal{U}(f, D) - \mathcal{U}(f_n, D)| + [\mathcal{U}(f_n, D) - \mathcal{L}(f_n, D)] + |\mathcal{L}(f_n, D) - \mathcal{L}(f, D)|$$
$$< \epsilon .$$

Hence, by Lemma 5.7, $f \in \mathcal{R}[a, b]$.

Denote $\int_a^b f$ by I. Let $\epsilon > 0$ be given. Then there exists a positive integer N such that $\|f_n - f\| < \epsilon/(b-a)$ for all $n > N$. It then follows that, for all $n > N$,

$$|I_n - I| = \left| \int_a^b (f_n - f) \right| \leq \int_a^b |f_n - f| \leq \|f_n - f\|(b-a) < \epsilon,$$

and so $(I_n) \to I$, as required.   □

One may remember this result as "The integral of the limit is the limit of the integrals". The following examples show that if the convergence of (f_n) is not uniform then the result may fail.

Example 7.7

Find a sequence of functions (f_n) in $\mathcal{R}[0, 1]$ whose pointwise limit is not Riemann integrable in $[0, 1]$.

Solution

We can lay out the rational numbers in $[0, 1]$ in a list

$$0, 1, \frac{1}{2}, \frac{1}{3}, \frac{2}{3}, \frac{1}{4}, \frac{3}{4}, \frac{1}{5}, \frac{2}{5}, \frac{3}{5}, \frac{4}{5}, \frac{1}{6}, \frac{5}{6}, \frac{1}{7}, \ldots .$$

(We are choosing denominators $1, 2, 3, \ldots$ in order, and then listing those irreducible fractions with that denominator in ascending order of numerator.) If we denote this sequence by $r_1, r_2, r_3, \ldots$, we can define functions f_n ($n \in \mathbb{N}$) by

$$f_n(x) = \begin{cases} 1 & \text{if } x \in \{r_1, r_2, \ldots, r_n\} \\ 0 & \text{otherwise.} \end{cases}$$

Since each f_n has only finitely many discontinuities, it is Riemann integrable. However, the pointwise limit of f_n is the function f given by

$$f(x) = \begin{cases} 1 & \text{if } x \text{ is rational} \\ 0 & \text{otherwise,} \end{cases}$$

since, for each x in $[0, 1]$, either x is irrational, in which case $f_n(x) = f(x) = 0$ for all n; or $x = r_N$ for some N, in which case $f_n(x) = f(x) = 1$ for all $n \geq N$. The function f is certainly not Riemann integrable in $[0, 1]$.

Observe that the sequence (f_n) does not have a uniform limit, since $\|f_m - f_n\| = 1$ for all $m \neq n$.   □

Example 7.8

For all $n \geq 2$, let

$$f_n(x) = \begin{cases} n^2 x & \text{if } 0 \leq x \leq 1/n \\ n(2 - nx) & \text{if } 1/n < x \leq 2/n \\ 0 & \text{if } 2/n < x \leq 1. \end{cases}$$

Show that each f_n is in $\mathcal{R}[0,1]$ and that the sequence (f_n) has a pointwise limit f in $\mathcal{R}[0,1]$, but that

$$\lim_{n \to \infty} \int_0^1 f_n \neq \int_0^1 f.$$

Solution

The graphs of f_2, f_4 and f_8 are shown in the following diagram:

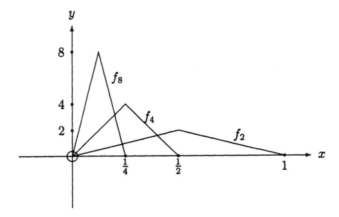

All the functions f_n are continuous (and so of course Riemann integrable). We now show that the pointwise limit of the sequence (f_n) is the zero function z given by $z(x) = 0$ $(x \in [0,1])$. To see this, observe first that $|f_n(0) - z(0)| = 0$ for all n. If $x \neq 0$ then there exists a natural number N such that $2/N < x$, and so $|f_n(x) - z(x)| = 0$ for all $n \geq N$. The function z is certainly Riemann integrable, with $\int_0^1 z = 0$. On the other hand, $\int_0^1 f_n$ is the area of a triangle with base $2/n$ and height n; that is, $\int_0^1 f_n = 1$ for all n. Hence

$$\lim_{n \to \infty} \int_0^1 f_n \neq \int_0^1 \left(\lim_{n \to \infty} f_n \right).$$

The sequence (f_n) does not have a uniform limit, since $\|f_n\| = \sup_{[0,1]} f_n = n$. $\qquad \square$

Although Theorem 7.6 is concerned with definite integrals, it is easy to modify it to deal with indefinite integrals. We have to be careful, of course, because of the very indefiniteness involved, but we can consider a function J_n, defined by

$$J_n(x) = \int_a^x f_n \quad (n \in \mathbb{N},\ x \in [a,b])\,. \tag{7.3}$$

Then $J_n' = f_n$, and so J_n is an indefinite integral of f_n. We define the function J by

$$J(x) = \int_a^x f \quad (x \in [a,b])\,; \tag{7.4}$$

then J is an indefinite integral of f.

Theorem 7.9

Let (f_n) be a sequence of functions in $\mathcal{R}[a,b]$, and suppose that $(f_n) \to f$ uniformly on $[a,b]$. Let J_n and J be as defined in (7.3) and (7.4). Then $(J_n) \to J$ uniformly in $[a,b]$.

Proof

Let $\epsilon > 0$ be given, and choose N so that $\|f_n - f\| < \epsilon/2(b-a)$ for all $n > N$. Then, for all $n > N$, and for all x in $[a,b]$,

$$|J_n(x) - J(x)| = \left| \int_a^x f_n - f \right| \leq \int_a^x |f_n - f| \leq (b-a)\|f_n - f\| < \frac{\epsilon}{2}\,.$$

It follows that

$$\|J_n - J\| = \sup\left\{ |J_n(x) - J(x)| : x \in [a,b] \right\} \leq \frac{\epsilon}{2} < \epsilon,$$

and so $(J_n) \to J$ uniformly in $[a,b]$, as required. $\square$

It is reasonable now to ask whether there is a corresponding theorem concerning differentiation of uniformly convergent sequences. The following example shows that our first guess as to what that theorem might be is certainly incorrect.

Example 7.10

Let

$$f_n(x) = \frac{1}{n} \sin nx \quad (n \geq 1,\ x \in [0, \pi/2])\,.$$

Show that $(f_n) \to 0$ uniformly in $[0, \pi/2]$, but that (f_n') is not uniformly convergent.

Solution

Since

$$\lim_{n \to \infty} \|f_n\| = \lim_{n \to \infty} \frac{1}{n} = 0 \,,$$

the sequence (f_n) tends uniformly to the zero function. However, except at $x = \pi/2$, the sequence (f'_n) does not have even a pointwise limit in $[0, \pi/2]$, since $f'_n(x) = \cos nx$.   □

In fact it is the uniform convergence of (f'_n), rather than the uniform convergence of (f_n), that is the key:

Theorem 7.11

Let (f_n) be a sequence of functions, differentiable on $[a, b]$, tending pointwise to f. Suppose that f'_n is continuous for all n, and that $(f'_n) \to g$ uniformly in $[a, b]$. Then $(f_n) \to f$ uniformly in $[a, b]$, and $g = f'$.

Proof

For each x in $[a, b]$, let

$$h_n(x) = \int_a^x f'_n \,.$$

Then $h'_n = f'_n$ since f'_n is continuous. Hence, for all x in $[a, b]$,

$$h_n(x) = f_n(x) + c_n$$

for some constant c_n. Since $h_n(a) = 0$, we deduce that $c_n = -f_n(a)$; thus

$$h_n(x) = f_n(x) - f_n(a) \,.$$

From Theorem 7.9 we know that $(h_n) \to h$ uniformly in $[a, b]$, where h is defined by

$$h(x) = \int_a^x g \quad (x \in [a, b]) \,.$$

Hence, for all x in $[a, b]$

$$f(x) - f(a) = \lim_{n \to \infty} (f_n(x) - f_n(a)) = \lim_{n \to \infty} h_n(x) = h(x) = \int_a^x g \,.$$

Since, by Theorem 7.5, g is continuous, it follows immediately by differentiation that $f' = g$.

Since $h_n \to h$ uniformly and since $f_n(x) = h_n(x) + f_n(a)$ and $f(x) = h(x) + f(a)$, it follows immediately that $(f_n) \to f$ uniformly.   □

EXERCISES

7.1 Suppose that $(f_n) \to f$ and $(g_n) \to g$ uniformly on $[a, b]$, and that f, g are bounded functions.

a) Show that $(f_n + g_n) \to f + g$ uniformly on $[a, b]$.

b) Show that $(f_n \cdot g_n) \to f \cdot g$ uniformly on $[a, b]$.

c) Suppose that $f_n(x)$ is non-zero for all x and for all n, and that there exists $\delta > 0$ with the property that $f(x) \geq \delta$ for all x in $[a, b]$. Show that $(1/f_n) \to 1/f$ uniformly in $[a, b]$.

7.2 Find a sequence of functions, each discontinuous at every point in $[0, 1]$, converging uniformly to a continuous function.

7.3 Let
$$f_n(x) = n^2 x^n (1 - x) \quad (n \in \mathbb{N},\ x \in [0, 1])\,.$$

Show that $(f_n) \to 0$ pointwise in $[0, 1]$, but that the convergence is not uniform. Show also that

$$\lim_{n \to \infty} \int_0^1 f_n \neq \int_0^1 \left(\lim_{n \to \infty} f_n(x) \right) dx\,.$$

7.4 Let
$$f_n(x) = n x^n (1 - x) \quad (n \in \mathbb{N},\ x \in [0, 1])\,.$$

Show that $(f_n) \to 0$ pointwise in $[0, 1]$, but that the convergence is not uniform. Show, however, that

$$\lim_{n \to \infty} \int_0^1 f_n = \int_0^1 \left(\lim_{n \to \infty} f_n(x) \right) dx\,.$$

7.5 Let
$$f_n(x) = x^n (1 - x) \quad (n \in \mathbb{N},\ x \in [0, 1])\,.$$

Show that $(f_n) \to 0$ uniformly in $[0, 1]$, but that (f_n') is not uniformly convergent.

7.6 It is easy to extend the definition of uniform convergence from the case of a closed finite interval to more general subsets of $\mathbb{R}$. Let $f_n : [0, \infty) \to \mathbb{R}$ and $f : [0, \infty) \to \mathbb{R}$ be given by

$$f_n(x) = \frac{x}{x + n}, \quad f(x) = 0\,.$$

a) Show that $(f_n) \to f$ pointwise in $[0, \infty)$.

b) Show that, for each $b > 0$, $(f_n) \to f$ uniformly in $[0, b]$.

c) Show that (f_n) does not converge uniformly to f in $[0, \infty)$.

7.7 Let $f_n : [0, \infty) \to \mathbb{R}$ and $f : [0, \infty) \to \mathbb{R}$ be given by

$$f_n(x) = \frac{nx}{1 + nx}, \quad f(x) = \begin{cases} 0 & \text{if } x = 0 \\ 1 & \text{otherwise.} \end{cases}$$

a) Show that $(f_n) \to f$ pointwise in $[0, \infty)$.

b) Show that, for all $b > 0$, $(f_n) \to f$ uniformly in $[b, \infty)$.

c) Show that (f_n) does not converge uniformly to f on $[0, \infty)$.

7.8 Let

$$f_n(x) = x + \frac{1}{n}, \quad f(x) = x \quad (n \in \mathbb{N}, \ x \in \mathbb{R}) .$$

Show that $(f_n) \to f$ uniformly in $\mathbb{R}$, but that f_n^2 (defined by $(f_n^2)(x) = (f(x))^2$) does not converge uniformly to f^2.

7.2 Uniform Convergence of Series

Let (f_n) be a sequence of functions with common domain $[a, b]$, and, for each $n \geq 1$, let

$$F_n = f_1 + f_2 + \cdots + f_n . \tag{7.5}$$

If the sequence (F_n) of sums converges uniformly on $[a, b]$ to a function F, then we say that the series $\sum_{n=1}^{\infty} f_n$ is **uniformly convergent** in $[a, b]$, and that it **converges**, or **sums**, **uniformly to** F. Similarly, if (F_n) converges pointwise on $[a, b]$ to F, then we say that the series $\sum_{n=1}^{\infty} f_n$ is **pointwise convergent** and that it **converges**, or **sums**, **pointwise to** F.

The following composite theorem is an easy consequence of the theorems in the previous section.

Theorem 7.12

Let $\sum_{n=1}^{\infty} f_n$ be a series of functions with common domain $[a, b]$.

(i) If $\sum_{n=1}^{\infty} f_n$ converges uniformly to F, then it converges pointwise to F.

(ii) If $\sum_{n=1}^{\infty} f_n$ converges uniformly to F in $[a, b]$, and if each f_n is continuous on $[a, b]$ then so is F.

(iii) If $\sum_{n=1}^{\infty} f_n$ sums uniformly to F in $[a, b]$, then

$$\sum_{n=1}^{\infty} \left(\int_a^b f_n \right) = \int_a^b F \left(= \int_a^b \left(\sum_{n=1}^{\infty} f_n \right) \right) .$$

(iv) If $\sum_{n=1}^{\infty} f_n$ sums uniformly to F in $[a, b]$, and if, for all $n \geq 1$, g_n is defined by

$$g_n(x) = \int_a^x f_n \quad (x \in [a, b]) ,$$

then $\sum_{n=1}^{\infty} g_n$ converges uniformly to G, where $G(x) = \int_a^x F \quad (x \in [a, b])$.

(v) If $\sum_{n=1}^{\infty} f_n$ sums pointwise to F in $[a, b]$, and if $\sum_{n=1}^{\infty} f_n'$ sums uniformly to G in $[a, b]$, then $\sum_{n=1}^{\infty} f_n$ sums uniformly to F, and $G = F'$.

Proof

(i) This is immediate from Theorem 7.4.

(ii) From Theorem 3.11, we have that F_n, as defined in (7.5), is continuous on $[a, b]$. The theorem then follows from Theorem 7.5.

(iii) From Theorem 5.15,

$$\int_a^b f_1 + \int_a^b f_1 + \cdots + \int_a^b f_n = \int_a^b F_n .$$

Hence, by Theorem 7.6,

$$\sum_{n=1}^{\infty} \left(\int_a^b f_n \right) = \lim_{n \to \infty} \int_a^b F_n = \int_a^b F .$$

(iv) For all $n \geq 1$ and for all x in $[a, b]$, let $G_n(x) = \sum_{r=1}^{n} g_r(x)$. We know from Theorem 5.15 that

$$G_n(x) = \int_a^x \left(\sum_{r=1}^{n} f_r \right) = \int_a^x F_n ,$$

and so, by Theorem 7.9, (G_n) converges uniformly to G, where, for all x in $[a, b]$, $G(x) = \int_a^x F$.

(v) For all $n \geq 1$, let $G_n = \sum_{r=1}^{n} f_n'$. Then $G_n = F_n'$ by Theorem 4.2. Since $G_n \to G$ uniformly in $[a, b]$, it follows from Theorem 7.11 that $(F_n) \to F$ uniformly in $[a, b]$, and that $G = F'$. $\qquad \square$

The processes legitimised by (iii) and (iv) are usually referred to as **integration term by term of a uniformly convergent series**, while the process in (v), which can be summarised as saying that

$$\sum_{n=1}^{\infty} f_n' = \left(\sum_{n=1}^{\infty} f_n \right)'$$

provided the series $\sum_{n=1}^{\infty} f_n'$ is uniformly convergent, is called **differentiation term by term**.

Example 7.13

Show that, for all x in $(-1, 1)$,

$$\log(1-x) = -x - \frac{1}{2}x^2 - \frac{1}{3}x^3 - \cdots = -\sum_{n=1}^{\infty} \frac{x^n}{n} . \qquad (7.6)$$

Solution

Let a be such that $0 \leq |x| < a < 1$, and consider the geometric series

$$1 + x + x^2 + \cdots \qquad (|x| \leq a) .$$

The sum to n terms is

$$G_n(x) = \frac{1 - x^n}{1 - x} ,$$

and the sum to infinity is

$$G(x) = \frac{1}{1 - x} .$$

Now, in the interval $[-a, a]$,

$$\begin{aligned}
\|G_n - G\| &= \sup \{|G_n(x) - G(x)| : x \in [-a, a]\} \\
&= \sup \{|x^n/(1 - x)| : x \in [-a, a]\} \\
&= a^n/(1 - a) \to 0
\end{aligned}$$

as $n \to \infty$. Hence $G_n \to G$ uniformly in $[-a, a]$. It follows that integration term by term is legitimate, and so, for all x in $[-a, a]$,

$$\int_0^x \frac{dt}{1 - t} = \int_0^x 1 \, dt + \int_0^x t \, dt + \int_0^x t^2 \, dt + \cdots .$$

That is,

$$-\log(1-x) = x + \frac{1}{2}x^2 + \frac{1}{3}x^3 + \cdots , \qquad (7.7)$$

as required.

Then, for example, putting $x = 1/2$ in (7.7) gives

$$\log 2 = \frac{1}{2} + \frac{1}{2}\left(\frac{1}{2}\right)^2 + \frac{1}{3}\left(\frac{1}{2}\right)^3 + \cdots .$$

$\square$

Example 7.14

Show that, for all x in $(-1, 1)$,

$$\tan^{-1} x = x - \frac{1}{3}x^3 + \frac{1}{5}x^5 - \cdots = \sum_{n=0}^{\infty} \frac{(-1)^n}{2n + 1}x^{2n+1} \qquad (x \in [-a, a]) . \qquad (7.8)$$

Solution

Let a be such that $0 \leq |x| < a < 1$. The geometric series

$$1 - t^2 + t^4 - \cdots \qquad (|t| \leq a) \tag{7.9}$$

has sum to n terms

$$F_n(t) = \frac{1 - (-t^2)^n}{1 + t^2},$$

and the sum to infinity is $F(t) = 1/(1 + t^2)$. Now,

$$\begin{aligned}
\|F_n - F\| &= \sup \{|F_n(t) - F(t)| : t \in [-a, a]\} \\
&= \sup \{t^{2n}/(1 + t^2) : t \in [-a, a]\} \\
&\leq a^{2n}.
\end{aligned}$$

Since $a^{2n} \to 0$ as $n \to \infty$, we conclude that the series (7.9) converges uniformly to $1/(1 + t^2)$ in $[-a, a]$. Hence, by Theorem 7.12(iv), for all x in $[-a, a]$,

$$\int_0^x \frac{dt}{1 + t^2} = \int_0^x 1 \, dt - \int_0^x t^2 \, dt + \int_0^x t^4 \, dt - \cdots .$$

That is,

$$\tan^{-1} x = x - \frac{1}{3} x^3 + \frac{1}{5} x^5 - \cdots,$$

as required. $\square$

The next example applies this result to obtain a series from which π may be calculated in a practical way.

Example 7.15

Show that

$$\tan^{-1} \frac{1}{2} + \tan^{-1} \frac{1}{3} = \frac{\pi}{4}$$

and hence obtain an infinite series whose sum is π.

Solution

Let $\alpha = \tan^{-1}(1/2)$, $\beta = \tan^{-1}(1/3)$; thus $\tan \alpha = 1/2$, $\tan \beta = 1/3$. From (3.6) we have

$$\begin{aligned}
\tan(\alpha + \beta) &= \frac{\sin(\alpha + \beta)}{\cos(\alpha + \beta)} = \frac{\sin \alpha \cos \beta + \cos \alpha \sin \beta}{\cos \alpha \cos \beta - \sin \alpha \sin \beta} \\
&= \frac{\tan \alpha + \tan \beta}{1 - \tan \alpha \tan \beta} = \frac{\frac{1}{2} + \frac{1}{3}}{1 - \frac{1}{2} \cdot \frac{1}{3}} = 1.
\end{aligned}$$

Since $\alpha, \beta \in (0, \pi/2)$, it follows that $\alpha + \beta \in (0, \pi)$. Thus $\alpha + \beta = \pi/4$, the unique angle in $(0, \pi)$ whose tangent is 1.

It follows that

$$\pi = 4\left[\left(\frac{1}{2}\right) - \frac{1}{3}\left(\frac{1}{2}\right)^3 + \frac{1}{5}\left(\frac{1}{2}\right)^5 - \cdots\right]$$
$$+ 4\left[\left(\frac{1}{3}\right) - \frac{1}{3}\left(\frac{1}{3}\right)^3 + \frac{1}{5}\left(\frac{1}{3}\right)^5 - \cdots\right].$$

$\square$

Remark 7.16

The terms inside both of the square brackets decrease fairly rapidly, and so a reasonably accurate estimate of π can be obtained without too much labour. A more rapidly converging series can be obtained using the formula

$$\frac{\pi}{4} = 4\tan^{-1}\frac{1}{5} - \tan^{-1}\frac{1}{239}.$$

There is no difficulty in principle in proving this equality, provided you have the patience to develop a formula for $\tan(4\alpha - \beta)$ and are not put off by the arithmetic involved.

As with the series of numbers encountered in Chapter 2, we need to be able to investigate the (uniform) convergence of a series of functions even if we have no formula for its sum to n terms. Here we appeal to a result known as the **general principle of uniform convergence**:

Theorem 7.17

Let (F_n) be a sequence of functions with common domain $[a, b]$. Then (F_n) is uniformly convergent in $[a, b]$ if and only if for every $\epsilon > 0$ there exists a positive integer N such that $\|F_m - F_n\| < \epsilon$ for all $m > n > N$.

Proof

Suppose first that (F_n) converges uniformly to F. Then for every $\epsilon > 0$ there exists N such that $\|F_n - F\| < \epsilon/2$ for all $n > N$. It follows that, for all $m > n > N$,

$$\|F_m - F_n\| = \|(F_m - F) + (F - F_n)\| \le \|F_m - F\| + \|F_n - F\| < \epsilon.$$

Conversely, suppose that for every $\epsilon > 0$ there exists N such that $\|F_m - F_n\| < \epsilon$ for all $m > n > N$. Let $\epsilon > 0$ be given. For technical reasons that will shortly be apparent, we choose N so that $\|F_m - F_n\| < \epsilon/2$ for all $m > n > N$. Then for each x in $[a, b]$ the number sequence $(F_n(x))$ has the property that

$$|F_m(x) - F_n(x)| < \frac{\epsilon}{2} < \epsilon$$

for all $m > n > N$, and so is a Cauchy sequence, as defined in Section 2.4. By Theorem 2.16, the sequence $(F_n(x))$ is convergent. For each x in $[a, b]$, let $\lim_{n \to \infty} F_n(x) = F(x)$.

From Exercise 2.24 we have that, for all $n > N$ and all x in $[a, b]$,

$$|F_n(x) - F(x)| \leq \frac{\epsilon}{2},$$

and from this it follows that $\|F_n - F\| \leq \epsilon/2 < \epsilon$ for all $n > N$. Thus $(F_n) \to F$ uniformly in $[a, b]$. $\qquad\square$

In terms of series, where $F_n = \sum_{r=1}^{n} f_r$, and where $F_m - F_n$ becomes $\sum_{r=n+1}^{m} f_r$, we have:

Theorem 7.18

Let (f_n) be a sequence of function with common domain $[a, b]$. Then $\sum_{n=1}^{\infty} f_n$ is uniformly convergent in $[a, b]$ if and only if for all $\epsilon > 0$ there exists a natural number N such that

$$\left\| \sum_{r=m+1}^{m} f_r \right\| < \epsilon$$

for all $m > n > N$. $\qquad\square$

The very useful **Weierstrass M-test** is an easy consequence of Theorem 7.18:

Theorem 7.19

Let (f_n) be a sequence of function with common domain $[a, b]$, and suppose that there exists a sequence (M_n) of real numbers with the property that $\|f_n\| \leq M_n$ for all $n \geq 1$. If $\sum_{n=1}^{\infty} M_n$ is convergent, then $\sum_{n=1}^{\infty} f_n$ is uniformly convergent in $[a, n]$.

Proof

Suppose that $\sum_{n=1}^{\infty} M_n$ is convergent. Then there is a natural number N with the property that, for all $m > n > N$,

$$\sum_{r=n+1}^{m} M_r < \epsilon.$$

It follows that

$$\left\| \sum_{r=n+1}^{m} f_r \right\| \leq \sum_{r=n+1}^{m} \|f_r\| \leq \sum_{r=n+1}^{m} M_r < \epsilon,$$

and so, by Theorem 7.18, $\sum_{n=1}^{\infty} f_n$ is uniformly convergent in $[a, b]$.   □

Example 7.20

Show that the series

$$\sum_{n=1}^{\infty} \frac{\cos nx}{n^2}$$

is uniformly convergent in every finite interval.

Solution

Since

$$\left| \frac{\cos nx}{n^2} \right| \leq \frac{1}{n^2}$$

for all x, and since $\sum_{n=1}^{\infty} (1/n^2)$ is convergent, it follows from Theorem 7.19 that the series is uniformly convergent in every finite interval.   □

We shall have occasion shortly to make use of **Abel's**[1] **test**:

Theorem 7.21

Suppose that

(i) $\sum_{n=1}^{\infty} f_n$ converges uniformly in $[a, b]$; and

(ii) for all x in $[a, b]$, $(g_n(x))$ is a decreasing sequence of positive numbers, and $g_1(x) \leq K$.

Then $\sum_{n=1}^{\infty} f_n \cdot g_n$ is uniformly convergent in $[a, b]$.

---

[1] Nils Henrik Abel, 1802–1829

Proof

Let $\epsilon > 0$ be given. There exists N such that, for all x in $[a, b]$ and for all $m > n > N$,

$$\left| \sum_{r=n+1}^{m} f_r(x) \right| < \frac{\epsilon}{K} \tag{7.10}$$

Let $n > N$. For all x in $[a, b]$ define $S_n(x) = 0$, and for all $m \geq n + 1$ let

$$S_m(x) = \sum_{r=n+1}^{m} f_r(x), \quad T_m(x) = \sum_{r=n+1}^{m} f_r(x) g_r(x).$$

Then, for all m, n such that $m > n > N$,

$$T_m(x) - T_n(x) = \sum_{r=n+1}^{m} f_r(x) g_r(x)$$

$$= \sum_{r=n+1}^{m} [S_r(x) - S_{r-1}(x)] g_r(x)$$

$$= [S_{n+1}(x) - S_n(x)] g_{n+1}(x) + [S_{n+2}(x) - S_{n+1}(x)] g_{n+2}(x) + \cdots$$

$$\cdots + [S_{m-1}(x) - S_{m-2}(x)] g_{m-1}(x) + [S_m(x) - S_{m-1}(x)] g_m(x)$$

$$= S_{n+1}(x)[g_{n+1}(x) - g_{n+2}(x)] + S_{n+2}(x)[g_{n+2}(x) - g_{n+3}(x)] + \cdots$$

$$+ S_{m-1}(x)[g_{m-1}(x) - g_m(x)] + S_m(x) g_m(x)$$

$$= \sum_{j=n+1}^{m-1} S_j(x)[g_j(x) - g_{j+1}(x)] + S_m(x) g_m(x).$$

Hence

$$|T_m(x) - T_n(x)| \leq \sum_{j=n+1}^{m-1} \frac{\epsilon}{K}[g_j(x) - g_{j+1}(x)] + \frac{\epsilon}{K} g_m(x)$$

$$= \frac{\epsilon}{K} g_{n+1}(x) \leq \frac{\epsilon}{K} g_1(x) < \epsilon.$$

Hence, by Theorem 7.18, $\sum_{n=1}^{\infty} f_n \cdot g_n$ is uniformly convergent in $[a, b]$. $\quad\square$

Example 7.22

Show that

$$\sum_{n=1}^{\infty} \frac{(-1)^n}{x + n}$$

is uniformly convergent in $[0, \infty)$. What about

$$\sum_{n=1}^{\infty} \frac{(-1)^n (1 + |\cos^n x|)}{x + n} \, ?$$

Solution

By the Leibniz test (Theorem 2.37) the series converges for each positive x. Denote the sum by $F(x)$ and the sum to n terms by $F_n(x)$. Then, for all x in $[0, \infty)$ and all N in $\mathbb{N}$,

$$F(x) - F_{2N-1}(x) = \sum_{n=2N}^{\infty} \frac{(-1)^n}{x+n} = \sum_{k=N}^{\infty} \left(\frac{1}{x+2k} - \frac{1}{x+2k+1} \right)$$

$$= \sum_{k=N}^{\infty} \frac{1}{(x+2k)(x+2k+1)} \leq \sum_{k=N}^{\infty} \frac{1}{2k(2k+1)}$$

$$< \sum_{n=2N}^{\infty} \frac{1}{n(n+1)} = \sum_{n=2N}^{\infty} \left(\frac{1}{n} - \frac{1}{n+1} \right)$$

$$= \frac{1}{2N},$$

and a similar argument shows that $0 \leq F_{2N}(x) - F(x) \leq 1/(2N+1)$. It is therefore possible to choose a positive integer M with the property that, for all x in $[0, \infty)$

$$|F(x) - F_n(x)| < \epsilon \text{ for all } n > M,$$

and so the series is uniformly convergent in $[0, \infty)$.

The sequence $(1 + |\cos^n x|)_{n \in \mathbb{N}}$ is decreasing, with first entry $1 + |\cos x| \leq 2$. Hence, by Abel's test, the series

$$\sum_{n=1}^{\infty} \frac{(-1)^n (1 + |\cos^n x|)}{x+n}$$

is uniformly convergent in $[0, \infty)$.   □

EXERCISES

7.9 Show that

$$\sum_{n=1}^{\infty} \frac{nx^2}{n^3 + x^3}$$

is uniformly convergent in any finite interval $[0, b]$.

7.10 Investigate pointwise and uniform convergence for the following series $\sum_{n=1}^{\infty} f_n$. Assume that $x \in [0, \infty)$. If there is uniform convergence only for a subset of $[0, \infty)$, find that subset.

$$f_n(x) = \frac{x^n}{x^n + 1}, \quad f_n(x) = \frac{1}{n^2(x+1)^2}, \quad f_n(x) = \frac{1}{x^n + 1}.$$

7.11 Investigate the pointwise and uniform convergence of the series

$$\sum_{n=0}^{\infty} \frac{x^2}{(1+x^2)^n} .$$

7.12 Determine whether

$$\sum_{n=1}^{\infty} \frac{x}{n^{3/2} + n^{3/4}x^2} \quad \text{and} \quad \sum_{n=1}^{\infty} \frac{x}{n^{3/4} + n^{3/2}x^2}$$

are uniformly convergent in $[0,1]$.

7.13 Investigate the pointwise and uniform convergence in $[0,1]$ of the series

$$\sum_{n=1}^{\infty} \frac{x^n(1-x)}{n^2} \quad \text{and} \quad \sum_{n=1}^{\infty} \frac{x^n(1-x)}{n} .$$

7.14 Show that

$$\sum_{n=1}^{\infty} \frac{1}{n} \left(\log(n+x) - \log n \right)$$

is uniformly convergent in $[0,1]$.

7.3 Power Series

The most important case of a series $\sum_{n=0}^{\infty} f_n$ occurs when f_n is the function $x \mapsto a_n x^n$ $(n \geq 0)$. This is what we call a **power series**

$$\sum_{n=0}^{\infty} a_n x^n . \tag{7.11}$$

For a given sequence (a_n) of coefficients, the convergence of the series depends on the value of x.

Theorem 7.23

If the series (7.11) is convergent for $x = R$, then, for every r such that $0 \leq r < |R|$, it is absolutely and uniformly convergent in the interval $[-r, r]$.

Proof

Since $\sum_{n=0}^{\infty} a_n R^n$ converges, we must have $|a_n R^n| \leq K$ for some $K > 0$. Since $|r/R| < 1$, the geometric series $\sum_{n=0}^{\infty} |r/R|^n$ is convergent. Now, for all $n \geq 0$ and all x in $[-r, r]$

$$|a_n x^n| = |a_n R^n| \left|\frac{x}{R}\right|^n \leq K \left|\frac{x}{R}\right|^n \leq K \left|\frac{r}{R}\right|^n,$$

and hence, by Theorem 7.19, $\sum_{n=0}^{\infty} |a_n x^n|$ is uniformly convergent in $[-r, r]$. The series $\sum_{n=0}^{\infty} a_n x^n$, being absolutely convergent, is convergent for all x in $[-r, r]$. To show that its convergence is uniform we use the uniform convergence of $\sum_{n=0}^{\infty} |a_n x^n|$, which tells us that for all $\epsilon > 0$ there exists a natural number N such that, for all x in $[-r, r]$ and all $m > n > N$,

$$\sum_{j=n+1}^{m} |a_j x^j| < \epsilon.$$

It follows that

$$\left|\sum_{j=n+1}^{m} a_j x^j\right| \leq \sum_{j=n+1}^{m} |a_j x^j| < \epsilon,$$

and so, by Theorem 7.18, $\sum_{n=0}^{\infty} a_n x^n$ converges uniformly in $[-r, r]$. $\square$

As a consequence of this result, we now have the crucial theorem concerning power series:

Theorem 7.24

A power series $\sum_{n=0}^{\infty} a_n x^n$ satisfies exactly one of the following three conditions:

 (i) the series converges for all x;

 (ii) the series converges only for $x = 0$;

 (iii) there exists a positive real number $R > 0$ such that the series converges for all x in $(-R, R)$ and diverges for all x in $(-\infty, -R) \cup (R, \infty)$.

Proof

Let $C = \{x \in [0, \infty) : \sum_{n=0}^{\infty} a_n x^n \text{ is convergent}\}$. If $C = [0, \infty)$ then we have Case (i). Otherwise C is bounded. If $C = \{0\}$ then we have Case (ii). Otherwise there is a positive real number $R = \sup C$. Suppose first that $y \in (-R, R)$. Then $(|y| + R)/2$ is *not* an upper bound for C, and so there exists $y' \geq (|y| + R)/2 > |y|$ such that $\sum_{n=0}^{\infty} a_n (y')^n$ is convergent. It follows from Theorem 7.23

that $\sum_{n=0}^{\infty} a_n x^n$ is convergent for all x in $(-y', y')$, and so in particular that $\sum_{n=0}^{\infty} a_n y^n$ is convergent.

Next, suppose that $|y| > R$, and suppose, for a contradiction, that $\sum_{n=0}^{\infty} a_n y^n$ is convergent. Then, by Theorem 7.23, $\sum_{n=0}^{\infty} a_n x^n$ converges for all x in $(-|y|, |y|)$. In particular it converges for $x = (|y| + R)/2 > R$, and we have the desired contradiction. $\qquad\square$

The number R is referred to as the **radius**[2] **of convergence** of the series. We absorb Cases (i) and (ii) into this definition by saying[3] that $R = \infty$ in Case (i) and $R = 0$ in Case (ii).

Theorem 7.24 is silent about whether or not the series converges for $x = \pm R$. This is deliberate, for it is not possible to state a general result. The following very simple example is instructive.

Example 7.25

Investigate the convergence of

$$\sum_{n=0}^{\infty} n x^n, \quad \sum_{n=0}^{\infty} \frac{x^n}{n+1}, \quad \sum_{n=0}^{\infty} \frac{x^n}{(n+1)^2}.$$

Solution

We apply the ratio test to the first series:

$$\frac{|(n+1)x^{n+1}|}{|nx^n|} = \frac{n+1}{n} |x| \to |x| \text{ as } n \to \infty.$$

Hence the series $\sum_{n=0}^{\infty} |nx^n|$ converges if $|x| < 1$ and diverges if $|x| > 1$. By Remark 2.33 we even have that $nx^n \to \infty$ as $n \to \infty$ when $|x| > 1$, and so certainly $\sum_{n=0}^{\infty} nx^n$ is divergent in these circumstances. We deduce that $R = 1$. When $R = 1$ the series is

$$0 + 1 + 2 + 3 + \cdots,$$

which is obviously divergent; when $R = -1$ the series is

$$0 - 1 + 2 - 3 + \cdots,$$

and this is again divergent. The **interval of convergence** is $(-R, R)$.

---

[2] The word "radius" is much more natural if we consider *complex* power series, when the region of convergence is not an interval but a circular disc in the complex plane.

[3] As ever, when we write $R = \infty$, we mean precisely that the series converges for all x, no more, no less.

If we similarly apply the ratio test to the second series we see that

$$\frac{|(n+1)x^{n+1}|}{|(n+2)x^n|} = \frac{n+1}{n+2}|x| \to |x| \text{ as } n \to \infty,$$

and so again $R = 1$. When $x = 1$ the series becomes

$$1 + \frac{1}{2} + \frac{1}{3} + \cdots,$$

which is divergent; when $x = -1$ the series becomes

$$1 - \frac{1}{2} + \frac{1}{3} - \cdots,$$

which is convergent. The interval of convergence is $[-R, R)$.

Once again we apply the ratio test, this time to the third series:

$$\frac{|(n+1)^2 x^{n+1}|}{|(n+2)^2 x^n|} = \frac{(n+1)^2}{(n+2)^2}|x| \to |x| \text{ as } n \to \infty,$$

and so yet again $R = 1$. When $x = 1$ the series becomes

$$1 + \frac{1}{2^2} + \frac{1}{3^2} + \cdots,$$

which is convergent; when $x = -1$ the series becomes

$$1 - \frac{1}{2^2} + \frac{1}{3^2} - \cdots,$$

which is again convergent. The interval of convergence is $[-R, R]$. $\square$

The use of the ratio test here gives the clue to one useful method of determining the radius of convergence of a power series:

Theorem 7.26

For the power series $\sum_{n=0}^{\infty} a_n x^n$, if

$$\lim_{n \to \infty} \left| \frac{a_n}{a_{n+1}} \right| = \lambda > 0,$$

then the radius of convergence is λ.

Proof

As in the example above, we apply the ratio test:

$$\frac{|a_{n+1} x^{n+1}|}{|a_n x^n|} = \frac{a_{n+1}}{a_n}|x| \to \frac{|x|}{\lambda} \text{ as } n \to \infty.$$

Hence the series is (absolutely) convergent when $|x| < \lambda$ and (in view of Remark 2.33) divergent when $|x| > \lambda$. It follows that λ is the radius of convergence. $\square$

Remark 7.27

This result holds good even when $\lambda = 0$ (when the series converges only for $x = 0$), or when $|(a_n/a_{n+1})| \to \infty$, when the series converges for all x. For example, if $a_n = n^n$, then

$$\lim_{n \to \infty} \left| \frac{a_n}{a_{n+1}} \right| = \lim_{n \to \infty} \frac{n^n}{(n+1)^{n+1}} = \left(1 + \frac{1}{n}\right)^{-n} \cdot \frac{1}{n+1} \to e^{-1}.0 = 0,$$

and so the series converges only for $x = 0$. By contrast, if $a_n = 1/n!$, then

$$\left| \frac{a_n}{a_{n+1}} \right| = \frac{(n+1)!}{n!} = n + 1 \to \infty \text{ as } n \to \infty,$$

and so the series converges for all x.

An alternative way of obtaining the radius of convergence is the **nth root test**:

Theorem 7.28

For the power series $\sum_{n=0}^{\infty} a_n x^n$, if

$$\lim_{n \to \infty} |a_n|^{-1/n} = \lambda,$$

then the radius of convergence is λ.

Proof

We show that the series converges when $|x| < \lambda$ and diverges when $|x| > \lambda$. Suppose first that $|x| < \lambda$, and let μ be such that $|x| < \mu < \lambda$. If we take $\epsilon = \lambda - \mu$ and choose N so that $\left| |a_n^{-1/n}| - \lambda \right| < \lambda - \mu$ for all $n > N$, then $|a_n|^{-1/n} > \mu$. Hence $|a_n| < 1/\mu^n$, and so, for all $n > N$,

$$|a_n x^n| < \left(\frac{x}{\mu}\right)^n .$$

It follows that the series converges, by comparison with the geometric series $\sum(|x|/\mu)^n$, the latter series being convergent, since $0 \leq |x|/\mu < 1$.

If $|x| > \lambda$ we argue in a very similar way. Let μ be such that $|x| > \mu > \lambda$, and choose N so that $|a_n|^{-1/n} < \mu$ for all $n > N$. Then

$$|a_n x^n| > \left(\frac{x}{\mu}\right)^n ,$$

and so the series diverges by comparison with the divergent geometric series $\sum(|x|/\mu)^n$. It follows that λ is the radius of convergence. $\qquad\square$

Remark 7.29

As with Theorem 7.26, this holds good even when $\lim_{n\to\infty} |a_n|^{-1/n} = 0$ or when $|a_n|^{-1/n} \to \infty$ as $n \to \infty$.

Example 7.30

Determine the interval of convergence of $\sum_{n=0}^{\infty} a_n x^n$, where

$$a_n = \left(\frac{n}{n+1}\right)^{n^2}.$$

Solution

We use Theorem 7.28: for all $n \geq 1$,

$$|a_n|^{-1/n} = \left(\frac{n+1}{n}\right)^n = \left(1 + \frac{1}{n}\right)^n \to e \text{ as } n \to \infty.$$

Hence $R = e$.

Let $|x| = e$; then

$$|a_n x^n| = \left(\frac{n}{n+1}\right)^{n^2} e^n,$$

and so

$$\log|a_n x^n| = n - n^2 \log\left(1 + \frac{1}{n}\right).$$

Now, from Exercise 6.4 we know that

$$\frac{1}{n} - \frac{1}{2n^2} < \log\left(1 + \frac{1}{n}\right) < \frac{1}{n} - \frac{1}{2n^2} + \frac{1}{3n^3}.$$

Hence

$$n - n^2\left(\frac{1}{n} - \frac{1}{2n^2} + \frac{1}{3n^3}\right) < \log|a_n x^n| < n - \left(\frac{1}{n} - \frac{1}{2n^2}\right).$$

That is,

$$\frac{1}{2} - \frac{1}{3n} < \log|a_n x^n| < \frac{1}{2},$$

and from this we deduce that $\lim_{n\to\infty} |a_n x^n| = 1/2$. Since the exponential function is continuous, it follows that, if $|x| = e$, then

$$\lim_{n\to\infty} |a_n x^n| = e^{1/2}.$$

Since this limit is non-zero, the series definitely does not converge for $x = \pm e$. So the interval of convergence is $(-e, e)$. $\qquad\square$

It is clear that we can use a power series to *define* a function, whose domain is the interval of convergence of the series. The following result is important in enabling us to work with functions defined in this way:

Theorem 7.31

The power series $\sum_{n=0}^{\infty} a_n x^n$ and $\sum_{n=1}^{\infty} n a_n x^{n-1}$ have the same radius of convergence.

Proof

Suppose that the series $\sum_{n=0}^{\infty} a_n x^n$ and $\sum_{n=1}^{\infty} n a_n x^{n-1}$ have radii of convergence R_1, R_2, respectively. For each $x \neq 0$,

$$|a_n x^n| \leq |x||n a_n x^{n-1}|.$$

Hence, by the comparison test, the series $\sum_{n=0}^{\infty} a_n x^n$ is absolutely convergent for every x with the property that $\sum_{n=1}^{\infty} n a_n x^{n-1}$ is absolutely convergent, that is, for every x such that $|x| < R_2$. It follows that $R_2 \leq R_1$.

Suppose now, for a contradiction, that $R_2 < R_1$, and let x_1, x_2 be such that

$$R_2 < |x_2| < |x_1| < R_1.$$

From Exercise 1.20 we have that

$$n \left| \frac{x_2}{x_1} \right|^{n-1} < \frac{|x_1|}{|x_1| - |x_2|},$$

and from this we deduce that, for all $n \geq 2$,

$$|n a_n x_2^{n-1}| < \frac{1}{|x_1| - |x_2|} |a_n x_1^n|.$$

Since $|x_1| < R_1$, the series $\sum_{n=0}^{\infty} |a_n x_1^n|$ is convergent. Hence, by the comparison test, $\sum_{n=1}^{\infty} |n a_n x_2^{n-1}|$ also converges, and this is a contradiction, since $|x_2| > R_2$. We deduce that $R_1 = R_2$. $\qquad \square$

Remark 7.32

The theorem holds good for series with zero or infinite radius of convergence.

From Theorem 7.31 we immediately deduce the following important result:

Theorem 7.33

For a power series $\sum_{n=0}^{\infty} a_n x^n$ with radius of convergence R, both the differentiated series $\sum_{n=1}^{\infty} n a_n x^{n-1}$ and the integrated series $\sum_{n=0}^{\infty} (1/(n+1)) a_n x^{n+1}$ also have radius of convergence R.

From Theorem 7.12 we now know that, for a function f defined by

$$f(x) = \sum_{n=0}^{\infty} a_n x^n,$$

a power series with radius of convergence R,

$$f'(x) = \sum_{n=1}^{\infty} n a_n x^{n-1} \quad (-R < x < R),$$

and, for all x in $(-R, R)$,

$$\int_0^x f(t)\, dt = \sum_{n=0}^{\infty} \frac{a_n}{n+1} x^{n+1}.$$

We have already encountered this process in Examples 7.13 and 7.14. In Example 7.13 the original series

$$1 + x + x^2 + \cdots$$

has interval of convergence $(-1, 1)$. The integrated series has the same radius of convergence, but we receive a small "bonus", in that the interval of convergence is $[-1, 1)$. The general theory of power series we have now developed certainly allows us to conclude that

$$x + \frac{1}{2}x^2 + \frac{1}{3}x^3 + \cdots = -\log(1-x) \tag{7.12}$$

whenever $|x| < 1$. In this case we know from Example 6.6 that (7.12) holds good for $x = -1$, but as yet we have no general theorem to tell us that this must be so.

In Example 7.14 we receive a double bonus from the integration process, since the integrated series

$$x - \frac{1}{3}x^3 + \frac{1}{5}x^5 - \cdots$$

has interval of convergence $[-1, 1]$, but our theory so far does not allow us to conclude that

$$1 - \frac{1}{3} + \frac{1}{5} - \cdots = \tan^{-1} 1 = \frac{\pi}{4}.$$

For this we need a result called **Abel's Theorem**:

Theorem 7.34

Consider a power series $\sum_{n=0}^{\infty} a_n x^n$ with radius of convergence R, and suppose that $\sum_{n=0}^{\infty} a_n x^n = s(x)$ for $-R < x < R$. Suppose further that $\sum_{n=0}^{\infty} a_n R^n$ is convergent. Then $\sum_{n=0}^{\infty} a_n R^n = \lim_{x \to R-} s(x)$.

Proof

We use Theorem 7.21 with $f_n(x) = a_n R^n$ for $n \geq 0$ and $0 \leq x \leq R$ (a constant function, but this creates no problem) and $g_n(x) = x^n/R^n$. We can certainly say that $\sum_{n=0}^{\infty} f_n(x)$ is uniformly convergent in $[0, R]$. It is also clear that $(g_n(x))$ is decreasing for all x in $[0, R]$, and that $g_1(x) \leq 1$. It therefore follows from Theorem 7.21 that

$$\sum_{n=0}^{\infty} a_n R^n \frac{x^n}{R^n} = \sum_{n=0}^{\infty} a_n x^n$$

is uniformly convergent in $[0, R]$. By Theorem 7.12, it now follows that the function $x \mapsto \sum_{n=0}^{\infty} a_n x^n$ is continuous in $[0, R]$, and so $\lim_{x \to R-} s(x) = \sum_{n=0}^{\infty} a_n R^n$. □

Remark 7.35

The theorem refers to convergence at R, but one can modify the proof to cope with convergence at $-R$, using the observation that the series $\sum_{n=0}^{\infty}(-1)^n a_n x^n$ converges in $[0, R]$ if and only if the series $\sum_{n=0}^{\infty} a_n x^n$ converges in $[-R, 0]$.

If we apply Theorem 7.34 to the series

$$x - \frac{1}{3}x^3 + \frac{1}{5}x^5 - \cdots ,$$

convergent in $[-1, 1]$, we even have *uniform* convergence in $[-1, 1]$, and

$$1 - \frac{1}{3} + \frac{1}{5} - \cdots = \lim_{x \to 1-} \tan^{-1} x = \frac{\pi}{4} . \qquad (7.13)$$

Putting $x = -1$ gives us essentially the same information:

$$-1 + \frac{1}{3} - \frac{1}{5} + \cdots = \lim_{x \to -1+} \tan^{-1} x = -\frac{\pi}{4} .$$

The series (7.13) is usually called **Gregory's**[4] **series**.

Our first encounter with power series was at the end of Chapter 4, when we mentioned Taylor–Maclaurin series. We now have the following quite comforting result:

---

[4] James Gregory, 1638–1675

Theorem 7.36

If $f(x)$ is defined by means of a power series $\sum_{n=0}^{\infty} a_n x^n$ with radius of convergence R, then

$$a_n = \frac{1}{n!} f^{(n)}(0) \quad (n \geq 0),$$

and so $\sum_{n=0}^{\infty} a_n x^n$ is the Taylor–Maclaurin series for f.

Proof

Certainly $a_0 = f(0)$. Applying Theorem 7.12 n times, we see that, for all x in $(-R, R)$,

$$f^{(n)}(x) = n! a_n + [(n+1)\ldots\ldots 2] a_{n+1} x + \text{higher powers of } x,$$

and so $f^{(n)}(0) = n! a_n$.   □

Remark 7.37

Let $f(x)$ be defined by a power series $\sum_{n=0}^{\infty} a_n x^n$ with radius of convergence R. It is a consequence of Theorem 7.36 that Maclaurin's Theorem (4.13) gives

$$f(x) = \sum_{n=0}^{N} a_n x^n + R_N.$$

Hence

$$R_N = \sum_{n=N+1}^{\infty} a_n x^n.$$

Since $\sum_{n=N+1}^{\infty} a_n x^n \to 0$ as $N \to \infty$ whenever $|x| < R$, we thus automatically see that $R_N \to 0$.

We have already come across some of the most important power series:

$$e^x = \sum_{n=0}^{\infty} \frac{x^n}{n!} \quad (x \in \mathbb{R}), \tag{7.14}$$

$$\log(1 + x) = \sum_{n=1}^{\infty} (-1)^{n-1} \frac{x^n}{n} \quad (-1 < x \leq 1), \tag{7.15}$$

$$\tan^{-1} x = \sum_{n=0}^{\infty} (-1)^n \frac{x^{2n+1}}{2n+1} \quad (-1 \leq x \leq 1) \tag{7.16}$$

and in the next chapter we will **define** the circular functions by series:

$$\cos x = = \sum_{n=0}^{\infty} (-1)^n \frac{x^{2n}}{(2n)!} = 1 - \frac{x^2}{2!} + \frac{x^4}{4!} - \cdots \quad (x \in \mathbb{R}) ; \qquad (7.17)$$

$$\sin x = \sum_{n=0}^{\infty} (-1)^n \frac{x^{2n+1}}{(2n+1)!} = x - \frac{x^3}{3!} + \frac{x^5}{5!} - \cdots \quad (x \in \mathbb{R}) . \qquad (7.18)$$

We conclude this chapter by examining the **binomial series**. First, it is useful to extend the definition of a binomial coefficient: for all integers $n > 0$ and all real numbers α, let

$$\binom{\alpha}{n} = \frac{\alpha(\alpha - 1) \cdots (\alpha - n + 1)}{n!} .$$

By convention we define $\binom{\alpha}{0}$ to be 1, and we observe that the Pascal triangle identity

$$\binom{\alpha - 1}{n} + \binom{\alpha - 1}{n + 1} = \binom{\alpha}{n + 1} \qquad (7.19)$$

holds good for a general α. If α is a positive integer then it is easy to see that $\binom{\alpha}{n} = 0$ for all $n \geq \alpha + 1$.

Let $f(x) = (1 + x)^\alpha$, where α is an arbitrary real number. Then certainly $f(0) = 1$. Next, by repeated differentiation, we find that, for all $n \geq 1$,

$$f^{(n)}(x) = \alpha(\alpha - 1) \cdots (\alpha - n + 1)(1 - x)^{\alpha - n} ,$$

and so

$$f^{(n)}(0) = n! \binom{\alpha}{n} .$$

Hence the Maclaurin series is

$$\sum_{n=0}^{\infty} \binom{\alpha}{n} x^n .$$

The radius of convergence of the series is found from Theorem 7.26:

$$\left| \frac{a_n}{a_{n+1}} \right| = \left| \frac{\alpha(\alpha - 1) \cdots (\alpha - n + 1)}{n!} \cdot \frac{(n + 1)!}{\alpha(\alpha - 1) \cdots (\alpha - n)} \right|$$

$$= \left| \frac{n + 1}{\alpha - n} \right| \to 1 \quad \text{as } n \to \infty.$$

Thus the radius of convergence is 1, and we have

$$(1 + x)^\alpha = \sum_{n=0}^{\infty} \binom{\alpha}{n} x^n \quad (-1 < x < 1) . \qquad (7.20)$$

Can we extend the interval to include 1 and -1? The quite complicated answer is given in the following theorem:

Theorem 7.38

(i) When $x = 1$, the binomial series (7.20) converges (with sum 2^α) if and only if $\alpha > -1$.

(ii) When $x = -1$, the series converges if and only if $\alpha \geq 0$. If $\alpha = 0$ the sum is 1; if $\alpha > 0$ the sum is 0.

Proof

Let $x = 1$, so that the series is

$$\sum_{n=0}^{\infty} \binom{\alpha}{n}, \qquad (7.21)$$

and suppose first that $\alpha \leq -1$. Then $\alpha - r \leq -r - 1$ for all r, and so $|\alpha - r| \geq r + 1$ for $r = 0, 1, \ldots, n - 1$. Hence

$$\left| \binom{\alpha}{n} \right| = \frac{|\alpha|}{1} \cdot \frac{|\alpha - 1|}{2} \cdots \frac{|\alpha - n + 1|}{n} \geq 1.$$

Since $\left(\left| \binom{\alpha}{n} \right| \right) \not\to 0$, the series (7.21) does not converge.

Next, suppose that $\alpha > -1$. If α is an integer the series terminates, and so convergence is automatic. So suppose that α is not an integer. For all $n > \alpha + 1$ the factor $\alpha - n + 1$ is negative, and so, from that point on, the terms of the series are alternately positive and negative in sign. From $\alpha > -1$ we deduce that $n - \alpha < n + 1$ for all n. Hence, denoting the terms of the series (7.21) by a_n for convenience, we have

$$|a_{n+1}| = \left| \frac{n - \alpha}{n + 1} \right| |a_n| < |a_n|,$$

and so $(|a_n|)$ is a strictly decreasing sequence. To deduce convergence from the Leibniz test (Theorem 2.37) we need to show that $(|a_n|) \to 0$. Now,

$$|a_n| = \left| (-1)^n \frac{(n - \alpha - 1)(n - \alpha - 2) \ldots (1 - \alpha)(-\alpha)}{n!} \right|$$

$$= \left| \frac{[n - (\alpha + 1)]}{n} \cdot \frac{[(n-1) - (\alpha + 1)]}{n - 1} \cdots \frac{[2 - (\alpha + 1)]}{2} \cdot \frac{[1 - (\alpha + 1)]}{1} \right|$$

$$= \left| \left(1 - \frac{\alpha + 1}{n} \right) \left(1 - \frac{\alpha + 1}{n - 1} \right) \cdots \left(1 - \frac{\alpha + 1}{2} \right) \left(1 - \frac{\alpha + 1}{1} \right) \right|.$$

Now recall the result of Exercise 6.3, that $1 - x \leq \exp{-x}$ for all x in $\mathbb{R}$. It follows from this that

$$|a_n| \leq \exp \left[(-\alpha + 1) \left(1 + \frac{1}{2} + \cdots + \frac{1}{n} \right) \right]$$

$$= \exp[-(\alpha + 1)(\log n + \gamma_n)], \quad \text{(from Example 6.6)}$$
$$= \exp[-(\alpha + 1)\log n]\exp[-(\alpha + 1)\gamma_n]$$
$$= n^{-(\alpha+1)}e^{-(\alpha+1)\gamma_n}.$$

Hence $(|a_n|) \to 0$, and the series (7.21) is convergent by the Leibniz test. By Abel's Theorem (Theorem 7.34), the sum of the series is 2^α.

Suppose now that $x = -1$. At this point we have a piece of good fortune, for we can easily show by induction that, for all $N \geq 1$

$$\sum_{n=0}^{N}(-1)^n\binom{\alpha}{n} = (-1)^N\binom{\alpha - 1}{N}.$$

The verification for $N = 1$ is trivial, and if we assume, for $N \geq 2$, that

$$\sum_{n=0}^{N-1}(-1)^n\binom{\alpha}{n} = (-1)^{N-1}\binom{\alpha - 1}{N - 1}$$

we deduce from (7.19) that

$$\sum_{n=0}^{N}(-1)^n\binom{\alpha}{n} = (-1)^N\left[\binom{\alpha}{N} - \binom{\alpha - 1}{N - 1}\right]$$
$$= (-1)^N\binom{\alpha - 1}{N}.$$

We have already observed that if $\alpha > 0$ then $\binom{\alpha-1}{N} \to 0$ as $N \to \infty$, and so

$$\sum_{n=0}^{\infty}(-1)^n\binom{\alpha}{n}$$

is convergent, with sum 0. If $\alpha = 0$ then convergence to the sum 1 is clear, since all terms after the first are zero.

So suppose that $\alpha < 0$, and write $\alpha = -\beta$, where $\beta > 0$. Using the same technique as before, we can show that

$$\left|(-1)^n\binom{\alpha - 1}{n}\right| = \left(1 - \frac{\alpha}{n}\right)\left(1 - \frac{\alpha}{n - 1}\right)\cdots\left(1 - \frac{\alpha}{1}\right)$$
$$= \left(1 + \frac{\beta}{n}\right)\left(1 + \frac{\beta}{n - 1}\right)\cdots\left(1 + \frac{\beta}{1}\right).$$

Hence

$$\log\binom{\alpha - 1}{n} = \log(1 + \beta) + \log\left(1 + \frac{\beta}{2}\right) + \cdots + \log\left(1 + \frac{\beta}{n}\right).$$

In Exercise 6.2 we saw that, for all $x > 0$

$$x - \frac{x^2}{2} \le \log(1+x) \le x.$$

Hence

$$\log \binom{\alpha-1}{n} \ge \beta \left(1 + \frac{1}{2} + \cdots + \frac{1}{n}\right) - \frac{\beta^2}{2}\left(1 + \frac{1}{2^2} + \cdots + \frac{1}{n^2}\right).$$

Since $\sum_{n=1}^{\infty}(1/n)$ diverges and $\sum_{n=1}^{\infty}(1/n^2)$ converges, this increases without limit as n increases. Hence $\binom{\alpha-1}{n} \to \infty$ as $n \to \infty$, and it follows that the series

$$\sum_{n=0}^{\infty}(-1)^n \binom{\alpha}{n}$$

is divergent when $\alpha < 0$.   □

Example 7.39

Find the Taylor–Maclaurin series for $\sin^{-1} x$.

Solution

Here there is no convenient formula for the nth derivative of the function, and we have to proceed in a more oblique manner. Denoting the function by f, we have

$$f'(x) = \frac{1}{\sqrt{1-x^2}}, \quad f''(x) = \frac{x}{(1-x^2)^{3/2}},$$

and so

$$(1-x^2)f''(x) - xf'(x) = 0.$$

We differentiate n times using Leibniz's Theorem (Theorem 4.19) and obtain

$$(1-x^2)f^{(n+2)}(x) + n(-2x)f^{(n+1)}(x) + \frac{n(n-1)}{2}(-2)f^{(n)}(x)$$

$$-xf^{(n+1)}(x) - n(1)f^{(n)}(x) = 0.$$

Putting $x = 0$ then gives us the "recurrence" equation

$$f^{(n+2)}(0) = n^2 f^{(n)}(0) \quad (n \ge 1).$$

Since $f'(0) = 1$ and $f''(0) = 0$, we deduce that $f^{(2n)}(0) = 0$ for all $n \ge 0$, and

$$f^{(2n+1)}(0) = 1^2.3^2.\ldots.(2n-1)^2 \quad (n \ge 0).$$

The Taylor–Maclaurin series is

$$\sum_{n=0}^{\infty} \frac{1^2.3^2.....(2n-1)^2}{(2n+1)!} x^{2n+1} .$$

The ratio of the magnitudes of successive terms is

$$\left| \frac{1^2.3^2.....(2n-1)^2}{(2n+1)!} \cdot \frac{(2n-1)!}{1^2.3^2.....(2n-3)^2} \right| \cdot x^2$$

$$= \frac{(2n-1)^2 x^2}{2n(2n+1)} \to x^2 \quad \text{as } n \to \infty,$$

and from this we deduce that the radius of convergence is 1.

To investigate the convergence at ± 1 we use **Stirling's formula**[5]

$$n! \sim \sqrt{2\pi n} \left(\frac{n}{e} \right)^n \quad \text{as } n \to \infty, \tag{7.22}$$

which we shall prove as formula (9.4) in Chapter 9. For our purposes here, the constant is irrelevant: all we need is that

$$n! \asymp n^{n+\frac{1}{2}} e^{-n} .$$

For $x = 1$ the series is

$$\sum_{n=0}^{\infty} \frac{1^2.3^2.....(2n-1)^2}{(2n+1)!} ,$$

and

$$\frac{1^2.3^2.....(2n-1)^2}{(2n+1)!} = \frac{[(2n-1)!]^2}{2^2.4^2.....(2n-2)^2(2n+1)!}$$

$$= \frac{(2n-1)!}{2n(2n+1)2^{2n-2}[(n-1)!]^2}$$

$$\asymp \frac{(2n-1)^{2n-\frac{1}{2}} e^{-(2n-1)}}{2n(2n+1)2^{2n-2}(n-1)^{2n-1} e^{-(2n-2)}}$$

$$= \left(\frac{2n-1}{2n-2} \right)^{2n-2} \frac{(2n-1)^{3/2} e}{2n(2n+1)(n-1)}$$

$$= \left(1 + \frac{1}{2n-2} \right)^{2n-2} \frac{(2n-1)^{3/2} e}{2n(2n+1)(n-1)}$$

$$\asymp n^{-3/2} \quad \text{as } n \to \infty,$$

since $\left[1 + (1/(2n-2)) \right]^{2n-2} \to e$. It follows that the series converges for $x = \pm 1$. □

---

[5] James Stirling, 1692–1770

EXERCISES

7.15 Find the interval of convergence of $\sum_{n=0}^{\infty} a_n x^n$, where

(i) $a_n = 2^n/(n+1)$;   (ii) $a_n = (-1)^n/\sqrt{n+1}$;

(iii) $a_n = n!/(n+1)^n$;   (iv) $a_n = 1/\left(2 + (1/(n+1))\right)^n$;

(v) $a_n = 1/(n+2)\log(n+2)$;   (vi) $a_n = (n!)^2/(2n)!$.

7.16 Show that, for all x in $(1,1)$,

$$\sum_{n=1}^{\infty} \frac{n}{n+1} x^{n+1} = \frac{1}{(1-x)} + \log(1-x) - 1\,.$$

7.17 Prove that $\sin 3x = 3\sin x - 4\sin^3 x$, and hence find the Taylor–Maclaurin series for $\sin^3 x$.

7.18 Suppose that the power series $\sum_{n=0}^{\infty} a_n x^n$ has radius of convergence R. Prove that, for all $|x| < \min\{1, R\}$,

$$\frac{1}{1-x} \sum_{n=0}^{\infty} a_n x^n = \sum_{n=0}^{\infty} s_n x^n\,,$$

where $s_n = a_0 + a_1 + \cdots + a_n$. Deduce that, for all $|x| < 1$,

$$\frac{1}{1-x}\log(1+x) = x + \left(1 - \frac{1}{2}\right) x^2 + \left(1 - \frac{1}{2} + \frac{1}{3}\right) x^3 + \cdots\,.$$

7.19 Let $f(x) = \sinh^{-1} x$ ($x \in \mathbb{R}$). Show that, for all x in $\mathbb{R}$

$$(1 + x^2)f''(x) + xf'(x) = 0\,.$$

By differentiating n times and equating x to 0, show that $f^{(n+2)}(0) = -n^2 f^{(n)}(0)$, and deduce that the Taylor–Maclaurin series for $\sinh^{-1}(x)$ is

$$\sum_{n=0}^{\infty} (-1)^{n+1} \frac{(2n)!}{(2n+1)\, 2^{2n}(n!)^2} x^{2n+1}\,.$$

The Circular Functions

8.1 Definitions and Elementary Properties

As was indicated in Chapter 3, we now give "official" analytic definitions of the circular functions cos and sin, and develop their properties. We **define**, for all x in $\mathbb{R}$,

$$\cos x = \sum_{n=0}^{\infty} \frac{x^{2n}}{(2n)!} = 1 - \frac{x^2}{2!} + \frac{x^4}{4!} - \cdots, \tag{8.1}$$

$$\sin x = \sum_{n=0}^{\infty} \frac{x^{2n+1}}{(2n+1)!} = x - \frac{x^3}{3!} + \frac{x^5}{5!} - \cdots. \tag{8.2}$$

From the definitions it is immediate that

$$\cos 0 = 1, \quad \sin 0 = 0, \tag{8.3}$$

and it is also clear that, for all x in $\mathbb{R}$,

$$\cos(-x) = \cos x, \quad \sin(-x) = -\sin x. \tag{8.4}$$

Both series converge for all x, and so both sin and cos are continuous for all x in $\mathbb{R}$. Also, differentiation term by term is a valid process, and we easily deduce that

$$(\sin)' x = \cos x, \quad (\cos)' x = -\sin x.$$

Let $F(x) = \cos^2 x + \sin^2 x$. Then

$$F'(x) = 2\cos x(-\sin x) + 2\sin x(\cos x) = 0$$

and so F is a constant function. From (8.3) we have that $F(0) = 1$, and so $F(x) = 1$ for all x in $\mathbb{R}$. Thus we have the identity

$$\cos^2 x + \sin^2 x = 1 \quad (x \in \mathbb{R}),$$

from which it follows that, for all x in $\mathbb{R}$,

$$0 \le |\cos x| \le 1, \quad 0 \le |\sin x| \le 1. \tag{8.5}$$

For a fixed y in $\mathbb{R}$, let

$$G_y(x) = [S_y(x)]^2 + [C_y(x)]^2,$$

where

$$S_y(x) = \sin(x + y) - \sin x \cos y - \cos x \sin y,$$
$$C_y(x) = \cos(x + y) - \cos x \cos y + \sin x \sin y.$$

It is easy to see that

$$S'_y(x) = C_y(x), \quad C'_y(x) = -S_y(x)$$

and so

$$\begin{aligned} G'_y(x) &= 2S_y(x)S'_y(x) + 2C_y(x)C'_y(x) \\ &= 2S_y(x)C_y(x) + 2C_y(x)\big(-S_y(x)\big) \\ &= 0. \end{aligned}$$

Hence the function G_y is constant. It is easy to see that $G_y(0) = 0$; hence $G_y(x) = 0$ for all x in $\mathbb{R}$. It follows that, for all x, y in $\mathbb{R}$,

$$\sin(x + y) = \sin x \cos y + \cos x \sin y, \tag{8.6}$$
$$\cos(x + y) = \cos x \cos y - \sin x \sin y. \tag{8.7}$$

Then, replacing y by $-y$, and using (8.4), we deduce

$$\sin(x - y) = \sin x \cos y - \cos x \sin y, \tag{8.8}$$
$$\cos(x - y) = \cos x \cos y + \sin x \sin y. \tag{8.9}$$

We know that $\cos 0 = 1$, and it might be that $\cos x > 0$ for all $x > 0$. Suppose, for a contradiction, that this is so. Since $\cos x = (\sin)' x$, it follows that $\sin$ is an increasing function. Thus $\sin 1 = k > 0$, and $\sin t \ge k$ for all $t \ge 1$. Let $x > 1$. Then, by the Mean Value Theorem (Theorem 4.7)

$$\cos x - \cos 1 = (x - 1)(-\sin t) \text{ for some } t \text{ in } (1, x)$$
$$\le -k(x - 1),$$

and so
$$\cos x \leq \cos 1 - k(x-1) \leq 0 \text{ when } x \geq ((\cos 1)/k) + 1.$$

From this contradiction we deduce that there exists $x > 0$ such that $\cos x = 0$. We now **define** the number π by the property that

$$\cos \frac{\pi}{2} = 0, \quad \cos x > 0 \text{ for all } x \text{ in } [0, \pi/2).$$

From (8.5) we then deduce that $\sin(\pi/2) = \pm 1$. Since $(\sin)' x = \cos x > 0$ in $[0, \pi/2)$, we must in fact have $\sin(\pi/2) = 1$. We can then use the addition formulae (8.6) and (8.7) to show

$$\sin \pi = 0, \ \cos \pi = -1, \ \sin 2\pi = 0, \ \cos 2\pi = 1.$$

Also,
$$\sin(\pi + x) = -\sin x, \ \cos(\pi + x) = -\cos x,$$
$$\sin(2\pi + x) = \sin x, \ \cos(2\pi + x) = \cos x.$$

The function sin is differentiable and has a positive derivative in $(-\pi/2, \pi/2)$, and so there exists an inverse function $\sin^{-1} : [-1, 1] \to [-\pi/2, \pi/2]$, differentiable in $(-1, 1)$. The function cos is differentiable and has a negative derivative in $(0, \pi)$ and so there exists an inverse function $\cos^{-1} : [-1, 1] \to [0, \pi]$, differentiable in $(-1, 1)$.

By Theorem 4.15,

$$(\sin^{-1})' x = \frac{1}{\cos(\sin^{-1} x)} = \frac{1}{\sqrt{[1 - [\sin(\sin^{-1} x)]^2]}} = \frac{1}{\sqrt{1 - x^2}}.$$

We choose the positive square root here, since $(\sin^{-1})'(x)$ is positive. By the same token, since $(\cos^{-1})' x$ is negative, we choose the negative square root in the following computation:

$$(\cos^{-1})' x = \frac{1}{-\sin(\cos^{-1} x)} = -\frac{1}{\sqrt{[1 - [\cos(\cos^{-1} x)]^2]}} = -\frac{1}{\sqrt{1 - x^2}}.$$

The differentiation properties of $\sin^{-1}$ and $\cos^{-1}$ give a hint as to the connection between the functions sin and cos and the geometry of the circle. But, as mentioned in Section 3.3, we need to be more precise about the meaning of the length of a curved line, and this is the issue that will be addressed in the next section.

8.2 Length

It will be convenient in this section to define curves by **parametric equations**. That is, a curve C is defined as

$$C = \{(r_1(t), r_2(t)) \; : \; t \in [a, b]\}$$

where $[a, b]$ is an interval, and r_1, r_2 are real functions with domain $[a, b]$. This has the advantage that there are no problems when the curve becomes vertical, or crosses itself:

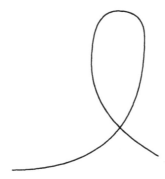

The most important example is the circle

$$x^2 + y^2 = a^2$$

with centre O and radius $a > 0$. The point $(a \cos t, a \sin t)$ $(t \in [0, 2\pi))$ certainly lies on the circle, by virtue of (8.5). Conversely, suppose that x, y are numbers such that $x^2 + y^2 = a^2$, and suppose first that x, y are both positive, so that (x, y) is in the first quadrant. Then $0 \le x/a \le 1$, and so, by the continuity and monotonicity of the function cos, there exists a unique t in $[0, \pi/2]$ such that $\cos t = x/a$. Thus $x = a \cos t$, and $y^2 = a^2(1 - \cos^2 t) = a^2 \sin^2 t$. Hence y, being positive, is equal to $a \sin t$.

If (x, y) is in the second quadrant, then $(-x, y)$, being in the first quadrant, is equal to $(a \cos t, a \sin t)$ for some t in $[0, \pi/2]$. Hence

$$(x, y) = \bigl(a \cos(\pi - t), a \sin(\pi - t)\bigr).$$

If (x, y) is in the third quadrant, then $(-x, -y)$, being in the first quadrant, is equal to $(a \cos t, a \sin t)$ for some t in $[0, \pi/2]$. Hence

$$(x, y) = \bigl(a \cos(\pi + t), a \sin(\pi + t)\bigr).$$

Finally, if (x, y) is in the fourth quadrant and $y \ne 0$, then $(x, -y)$, being in the first quadrant, is equal to $(a \cos t, a \sin t)$ for some t in $(0, \pi/2]$. Hence

$$(x, y) = \bigl(a \cos(2\pi - t), a \sin(2\pi - t)\bigr).$$

We conclude that

$$\{(x,y) \in \mathbb{R}^2 \ : \ x^2 + y^2 = a^2\} = \{(a\cos t, a\sin t) \ : \ a \in [0, 2\pi)\} \,.$$

Returning now to the general case, let us consider a curve

$$\mathcal{C} = \{(r_1(t), r_2(t)) \ : \ t \in [a, b]\} \,,$$

and let $D = \{a = t_0, t_1, \ldots, t_n = b\}$ be a dissection of $[a, b]$, as defined in Section 5.1, with $t_0 < t_1 < \cdots < t_n$. Each t_i in D corresponds to a point $P_i = (r_1(t_i), r_2(t_i))$ on the curve $\mathcal{C}$, and it is reasonable to estimate the length of curve $\mathcal{C}$ between the point $A = P_0$ and $B = P_n$ as

$$\mathcal{P}(\mathcal{C}, D) = |P_0 P_1| + |P_1 P_2| + \cdots + |P_{n-1} P_n| \,. \tag{8.10}$$

To express this in analytic terms we find it useful to use vector notation, and it may be necessary to remind ourselves of some basic vector properties and notations. We shall confine ourselves to vectors in two dimensions. First, if $\mathbf{a} = (a_1, a_2)$ and $\mathbf{b} = (b_1, b_2)$, then the **scalar** (or **inner**) product $\mathbf{a}.\mathbf{b}$ is defined by

$$\mathbf{a}.\mathbf{b} = a_1 b_1 + a_2 b_2 \,,$$

and the **norm** $\|\mathbf{a}\|$ is defined by

$$\|\mathbf{a}\| = \sqrt{\mathbf{a}.\mathbf{a}} = \sqrt{a_1^2 + a_2^2} \,.$$

The **Cauchy–Schwarz inequality** (1.25) states that, for all vectors $\mathbf{a}$ and $\mathbf{b}$,

$$|\mathbf{a}.\mathbf{b}| \le \|\mathbf{a}\| \, \|\mathbf{b}\| \,. \tag{8.11}$$

We shall denote the point $(r_1(t), r_2(t))$ by $\mathbf{r}(t)$, and write

$$|P_{i-1} P_i| = \|\mathbf{r}(t_i) - \mathbf{r}(t_{i-1})\| = \left[\left(r_1(t_i) - r_1(t_{i-1})\right)^2 + \left(r_2(t_i) - r_2(t_{i-1})\right)^2 \right]^{1/2} \,.$$

Thus the analytic version of (8.10) is

$$\mathcal{P}(\mathcal{C}, D) = \sum_{i=1}^{n} \|\mathbf{r}(t_i) - \mathbf{r}(t_{i-1})\| \,. \tag{8.12}$$

It is clear that if we refine the dissection D by adding extra points then $\mathcal{P}(\mathcal{C}, D)$ increases: if Q is a point between P_{i-1} and P_i, then, by the triangle inequality, the combined length of segments $P_{i-1}Q$ and QP_i is not less than the length of the segment $P_{i-1}P_i$.

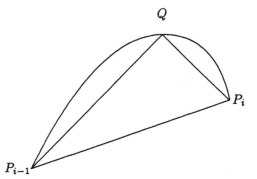

Let $\mathcal{D}$ be the set of all dissections of $[a, b]$. If $\{\mathcal{P}(\mathcal{C}, D) \ : \ D \in \mathcal{D}\}$ is bounded above, we say that the curve $\mathcal{C}$ is **rectifiable**, and we define its **length** $\Lambda(\mathcal{C})$ by

$$\Lambda(\mathcal{C}) = \sup\{\mathcal{P}(\mathcal{C}, D) \ : \ D \in \mathcal{D}\}.$$

Not every curve is rectifiable. To find an example of a curve that is not rectifiable we look again at a (continuous) function we have considered several times already:

Example 8.1

Let $\mathcal{C} = \{(t, r_2(t)) \ : \ t \in [0, 1]\}$, where

$$r_2(t) = \begin{cases} t\sin(1/t) & \text{if } t \neq 0 \\ 0 & \text{if } t = 0. \end{cases}$$

Show that $\mathcal{C}$ is not rectifiable.

Solution

For $n = 1, 2, 3, \ldots$, let

$$D_n = \left\{0, \frac{2}{n\pi}, \frac{2}{(n-1)\pi}, \ldots, \frac{2}{2\pi}, \frac{2}{\pi}, 1\right\}.$$

Observe that

$$r_2\left(\frac{2}{k\pi}\right) = \frac{2}{k\pi}\sin\left(\frac{k\pi}{2}\right) = \begin{cases} 0 & \text{if } k \text{ is even} \\ \pm 2/k\pi & \text{if } k \text{ is odd.} \end{cases}$$

Hence, if k is even,

$$\left\|\mathbf{r}\left(\frac{2}{k\pi}\right) - \mathbf{r}\left(\frac{2}{(k+1)\pi}\right)\right\| = \left\|\left(\frac{2}{k\pi}, 0\right) - \left(\frac{2}{(k+1)\pi}, \pm\frac{2}{(k+1)\pi}\right)\right\|$$

$$> \frac{2}{(k+1)\pi},$$

and if k is odd we can similarly show that

$$\left\| \mathbf{r}\left(\frac{2}{k\pi}\right) - \mathbf{r}\left(\frac{2}{(k+1)\pi}\right) \right\| > \frac{2}{k\pi} > \frac{2}{(k+1)\pi}.$$

It follows that

$$P(\mathcal{C}, D_n) > \frac{2}{\pi}\left(\frac{1}{2} + \frac{1}{3} + \cdots \frac{1}{n}\right),$$

and from the divergence of the harmonic series we see that there is no upper bound on the set $\{\mathcal{P}(\mathcal{C}, D) : D \in \mathcal{D}\}$. □

If $\mathbf{r}(t) = (r_1(t), r_2(t))$, and if r_1, r_2 are differentiable, then we define

$$\mathbf{r}'(t) = (r_1'(t), r_2'(t)), \quad \int_a^b \mathbf{r}(t)\, dt = \left(\int_a^b r_1(t)\, dt, \int_a^b r_2(t)\, dt\right).$$

We shall want to make use of the formula

$$\mathbf{c} \cdot \int_a^b \mathbf{r}(t)\, dt = \int_a^b \mathbf{c} \cdot \mathbf{r}(t)\, dt, \tag{8.13}$$

where $\mathbf{c} = (c_1, c_2)$ is a constant vector. The proof is easy:

$$\mathbf{c} \cdot \int_a^b \mathbf{r}(t)\, dt = c_1 \int_a^b r_1(t)\, dt + c_2 \int_a^b r_2(t)\, dt = \int_a^b [c_1 r_1(t) + c_2 r_2(t)]\, dt$$

$$= \int_a^b \mathbf{c} \cdot \mathbf{r}(t)\, dt.$$

We use this to prove the important inequality

$$\left\| \int_a^b \mathbf{r}(t)\, dt \right\| \le \int_a^b \|\mathbf{r}(t)\|\, dt. \tag{8.14}$$

To see that this is so, denote the constant vector $\int_a^b \mathbf{r}(t)\, dt$ by $\mathbf{c}$, and observe that the result is immediate if $\mathbf{c} = \mathbf{0}$. If $\mathbf{c} \ne \mathbf{0}$ we have that

$$\|\mathbf{c}\|^2 = \mathbf{c} \cdot \int_a^b \mathbf{r}(t)\, dt = \int_a^b \mathbf{c} \cdot \mathbf{r}(t)\, dt \quad \text{by (8.13)}$$

$$\le \int_a^b |\mathbf{c} . \mathbf{r}(t)|\, dt \quad \text{by Theorem 5.15}$$

$$\le \int_a^b \|\mathbf{c}\|\, \|\mathbf{r}(t)\|\, dt \quad \text{by (8.11)}$$

$$= \|\mathbf{c}\| \int_a^b \|\mathbf{r}(t)\|\, dt.$$

Then dividing by $\|\mathbf{c}\|$ gives the required result.

Returning now to the main theme of this section, we prove the following result:

Theorem 8.2

Let $C = \{\mathbf{r}(t) : t \in [a, b]\}$, where $\mathbf{r}(t) = (r_1(t), r_2(t))$, and suppose that r_1, r_2 are differentiable and r_1', r_2' are continuous on $[a, b]$. Then C is rectifiable, and

$$\Lambda(C) \leq \int_a^b \|\mathbf{r}'(t)\|\, dt. \tag{8.15}$$

Proof

For each dissection $D = \{a = t_0, t_1, \ldots, t_n = b\}$ of $[a, b]$,

$$
\begin{aligned}
P(C, D) &= \sum_{i=1}^{n} \|\mathbf{r}(t_i) - \mathbf{r}(t_{i-1})\| \\
&= \sum_{i=1}^{n} \left\| \int_{t_{i-1}}^{t_i} \mathbf{r}'(t)\, dt \right\| \quad \text{by the Fundamental Theorem} \\
&\leq \sum_{i=1}^{n} \int_{t_{i-1}}^{t_i} \|\mathbf{r}'(t)\|\, dt \quad \text{by (8.14)} \\
&= \int_a^b \|\mathbf{r}'(t)\|\, dt.
\end{aligned}
$$

$\square$

In fact we shall show that (8.15) is an equality. We begin by proving a not very surprising **additivity theorem**:

Theorem 8.3

Let
$$C = \{\mathbf{r}(t) : a \leq t \leq b\}$$
be a rectifiable curve, and let $c \in (a, b)$. If

$$C_1 = \{\mathbf{r}(t) : a \leq t \leq c\} \text{ and } C_2 = \{\mathbf{r}(t) : c \leq t \leq b\},$$

then C_1 and C_2 are both rectifiable, and

$$\Lambda(C) = \Lambda(C_1) + \Lambda(C_2).$$

Proof

Let D_1, D_2 be arbitrary dissections of $[a, c]$, $[c, b]$, respectively, and let $D = D_1 \cup D_2$. Then

$$P(C_1, D_1) + P(C_2, D_2) = P(C, D) \leq \Lambda(C). \tag{8.16}$$

Hence $P(\mathcal{C}_1, D_1) \leq \Lambda(\mathcal{C})$ and $P(\mathcal{C}_2, D_2) \leq \Lambda(\mathcal{C})$, and so both $\mathcal{C}_1$ and $\mathcal{C}_2$ are rectifiable. Also, from (8.16) we deduce that, for every dissection D_1 of $[a, c]$,

$$P(\mathcal{C}_1, D_1) \leq \Lambda(\mathcal{C}) - P(\mathcal{C}_2, D_2) ,$$

and so

$$\Lambda(\mathcal{C}_1) \leq \Lambda(\mathcal{C}) - P(\mathcal{C}_2, D_2) .$$

It follows that, for every dissection D_2 of $[c, b]$,

$$P(\mathcal{C}_2, D_2) \leq \Lambda(\mathcal{C}) - \Lambda(\mathcal{C}_1) ,$$

and so

$$\Lambda(\mathcal{C}_2) \leq \Lambda(\mathcal{C}) - \Lambda(\mathcal{C}_1) .$$

We have shown that

$$\Lambda(\mathcal{C}) \geq \Lambda(\mathcal{C}_1) + \Lambda(\mathcal{C}_2) .$$

To prove the opposite inequality, let D be an arbitrary dissection of $[a, b]$, and let $D' = D \cup \{c\}$. (It may of course turn out that $D' = D$, but this creates no problem.) Let $D_1 = D' \cap [a, c]$ and $D_2 = D' \cap [c, b]$. Then

$$P(\mathcal{C}, D) \leq P(\mathcal{C}, D') = P(\mathcal{C}_1, D_1) + P(\mathcal{C}_2, D_2) \leq \Lambda(\mathcal{C}_1) + \Lambda(\mathcal{C}_2) .$$

Since this holds for all dissections D, we deduce, as required, that

$$\Lambda(\mathcal{C}) \leq \Lambda(\mathcal{C}_1) + \Lambda(\mathcal{C}_2) .$$

$\square$

Let

$$\mathcal{C} = \{\mathbf{r}(t) : a \leq t \leq b\}$$

be a rectifiable curve, and for each t in $[a, b]$ let us denote the curve

$$\{\mathbf{r}(u) : a \leq u \leq t\}$$

by $\mathcal{C}_t$. Let s, the **arc-length function**, be defined by

$$s(t) = \Lambda(\mathcal{C}_t) \quad (t \in [a, b]) . \tag{8.17}$$

Theorem 8.4

Let

$$\mathcal{C} = \{\mathbf{r}(t) : a \leq t \leq b\} ,$$

and suppose that each of the components r_1, r_2 of $\mathbf{r}$ is differentiable with continuous derivative in $[a, b]$. Then the function s defined by (8.17) is monotonic increasing and differentiable, and

$$s'(t) = \|\mathbf{r}'(t)\| \quad (t \in [a, b]) .$$

Proof

Suppose that $a \leq t_1 \leq t_2 \leq b$. Let

$$C' = \{\mathbf{r}(u) : t_1 \leq u \leq t_2\}.$$

Then, by Theorem 8.3,

$$s(t_2) = s(t_1) + \Lambda(C') \geq s(t_1).$$

Let

$$f(t) = \int_a^t \|\mathbf{r}'(u)\| \, du \quad (t \in [a, b]).$$

Since $u \mapsto \|\mathbf{r}'(u)\|$ is continuous,

$$f'(t) = \|\mathbf{r}'(t)\| \quad (t \in [a, b]).$$

Let $h > 0$. We may think of the line joining $\mathbf{r}(t)$ and $\mathbf{r}(t+h)$ as a very simple approximation to the length of the curve

$$C_{t,h} = \{\mathbf{r}(u) : t \leq u \leq t+h\},$$

and so

$$\|\mathbf{r}(t+h) - \mathbf{r}(t)\| \leq \Lambda(C_{t,h}) = s(t+h) - s(t).$$

Hence, for all t, $t + h$ in $[a, b]$,

$$\left\| \frac{\mathbf{r}(t+h) - \mathbf{r}(t)}{h} \right\| \leq \frac{s(t+h) - s(t)}{h}$$

$$\leq \frac{1}{h} \int_t^{t+h} \|\mathbf{r}'(u)\| \, du \quad \text{(by Theorem 8.2)}$$

$$= \frac{f(t+h) - f(t)}{h}.$$

When $h < 0$ the details of the argument are a little different, but the conclusion, that

$$\left\| \frac{\mathbf{r}(t+h) - \mathbf{r}(t)}{h} \right\| \leq \frac{s(t+h) - s(t)}{h} \leq \frac{f(t+h) - f(t)}{h}, \tag{8.18}$$

still holds good.

Now, for all t in $[a, b]$,

$$\lim_{h \to 0} \left[\frac{f(t+h) - f(t)}{h} \right] = f'(t) = \|\mathbf{r}'(t)\|. \tag{8.19}$$

Also,

$$\left\| \frac{\mathbf{r}(t+h) - \mathbf{r}(t)}{h} \right\|^2 = \left[\frac{r_1(t+h) - r_1(t)}{h} \right]^2 + \left[\frac{r_2(t+h) - r_2(t)}{h} \right]^2$$

$$\to [r_1'(t)]^2 + [r_2'(t)]^2 = \|\mathbf{r}'(t)\|^2 \quad \text{as } h \to 0.$$

Hence

$$\left\| \frac{\mathbf{r}(t+h) - \mathbf{r}(t)}{h} \right\| \to \|\mathbf{r}'(t)\| \text{ as } h \to 0. \tag{8.20}$$

From (8.18), (8.19) and (8.20) it now follows that $s'(t) = \|\mathbf{r}'(t)\|$. $\qquad\square$

It is now an immediate consequence that, for all t in $[a, b]$,

$$s(t) = \int_a^t s'(u)\, du = \int_a^t \|\mathbf{r}'(u)\|\, du\,,$$

and in particular, that

$$\Lambda(\mathcal{C}) = s(b) = \int_a^b \|\mathbf{r}'(t)\|\, dt\,.$$

We state this conclusion formally as a theorem:

Theorem 8.5

Let $\mathcal{C} = \{\mathbf{r}(t) : t \in [a, b]\}$, where $\mathbf{r}(t) = (r_1(t), r_2(t))$, and suppose that r_1, r_2 are differentiable and r_1', r_2' are continuous on $[a, b]$. Then $\mathcal{C}$ is rectifiable, and

$$\Lambda(\mathcal{C}) = \int_a^b \|\mathbf{r}'(t)\|\, dt\,.$$

$\qquad\square$

Example 8.6

Find the length of the parabola $\mathcal{P} = \{(at^2, 2at) : t \in \mathbb{R}\}$. between the points $(a, -2a)$ and $(a, 2a)$.

Solution

The points in question correspond to $t = -1$ and $t = 1$. The vector $\mathbf{r}'(t)$ is $(2at, 2a)$, with norm $2a\sqrt{1 + t^2}$, and so the required length is

$$2a \int_{-1}^1 \sqrt{1 + t^2}\, dt\,.$$

Let

$$I = \int_{-1}^1 \sqrt{1 + t^2}\, dt\,.$$

In many cases the integral we obtain will be one that cannot be evaluated by elementary methods. Here we can actually perform the integration. The details are unimportant here, but the conclusion is that the length of the parabolic arc is

$$a\left(2\sqrt{2} + \log(\sqrt{2}+1) - \log(\sqrt{2}-1)\right).$$

□

The key application of Theorem 8.5 is to the unit circle

$$S = \{(\cos t, \sin t) : t \in [0, 2\pi)\}.$$

Here $\mathbf{r}'(t) = (-\sin t, \cos t)$, and so

$$\|\mathbf{r}'(t)\| = \sqrt{(\sin^2 t + \cos^2 t)} = 1.$$

Hence

$$s(t) = \int_0^t 1\, du = t.$$

Thus the parameter t measures the length of the arc of the unit circle from the point $(1, 0)$ to the point $(\cos t, \sin t)$. To put it another way, if the length of the arc to the point (x, y) on the circle is t, then $x = \cos t$ and $y = \sin t$. Thus the analytic and geometric definitions give rise to precisely the same functions.

We remark finally that the length of the unit circle from $t = 0$ to $t = 2\pi$ is 2π, and so our analytic definition of π gives precisely the same number as the standard geometric definition of π as the ratio of the circumference to the diameter of a circle.

EXERCISES

8.1 As an alternative approach to defining sin and cos, consider the function A defined by

$$A(x) = \int_0^x \frac{dt}{\sqrt{1-t^2}} \quad (-1 \le x \le 1).$$

a) Show that $A(0) = 0$, $A(-x) = -A(x)$, and that A is a strictly increasing function, differentiable in $(-1, 1)$, with

$$A'(x) = \frac{1}{\sqrt{1-x^2}}.$$

b) Define π to be $2A(1)$, and denote the inverse function A^{-1} : $[-\pi/2, \pi/2] \to [-1, 1]$ by S. Show that $S(0) = 0$, $S'(0) = 1$,

$$[S(x)]^2 + [S'(x)]^2 = 1,$$

and $S''(x) = -S(x)$.

9
Miscellaneous Examples

9.1 Wallis's Formula

Historically, the interest of Wallis's formula,

$$\lim_{n \to \infty} \frac{2^{2n}(n!)^2}{(2n)!\sqrt{n}} = \sqrt{\pi}, \tag{9.1}$$

is that it is one of the early examples of an arithmetical formula for π. In fact, although it looks so unexpected, it is not hard to prove.

Let

$$I_n = \int_0^{\pi/2} \sin^n x \, dx \quad (n \geq 0).$$

It is easy to calculate that $I_0 = \pi/2$ and $I_1 = 1$, and integration by parts leads to the recurrence formula

$$I_n = \frac{n-1}{n} I_{n-2}. \tag{9.2}$$

Repeated application of (9.2) leads to the formulae, encountered before in (5.21) and (5.22),

$$I_{2m} = \frac{2m-1}{2m} \cdot \frac{2m-3}{2m-2} \cdot \ldots \cdot \frac{1}{2} \cdot \frac{\pi}{2}, \quad I_{2m+1} = \frac{2m}{2m+1} \cdot \frac{2m-2}{2m-1} \cdot \ldots \cdot \frac{2}{3},$$

and by division we then obtain that, for all $m \geq 0$,

$$\frac{\pi}{2} = \frac{2.2}{1.3} \cdot \frac{4.4}{3.5} \cdot \ldots \cdot \frac{2m.2m}{(2m-1)(2m+1)} \cdot \frac{I_{2m}}{I_{2m+1}}. \tag{9.3}$$

We now show that $\lim_{m\to\infty}(I_{2m}/I_{2m+1}) = 1$. Since $0 < \sin x < 1$ in the interval $(0, \pi/2)$, we may deduce that

$$0 < \sin^{2m+1} x \le \sin^{2m} x \le \sin^{2m-1} x,$$

and hence that

$$0 < I_{2m+1} \le I_{2m} \le I_{2m+1}.$$

Hence, by (9.2)

$$1 \le \frac{I_{2m}}{I_{2m+1}} \le \frac{I_{2m-1}}{I_{2m+1}} = 1 + \frac{1}{2m},$$

and from this it is immediate that $\lim_{m\to\infty}(I_{2m}/I_{2m+1}) = 1$.

It follows that, in (9.3), we can let m tend to infinity and obtain the simplest version of Wallis's formula:

$$\frac{\pi}{2} = \lim_{m\to\infty} \left(\frac{2.2}{1.3} \cdot \frac{4.4}{3.5} \cdot \ldots \cdot \frac{2m.2m}{(2m-1)(2m+1)} \right).$$

Hence, since $\lim_{m\to\infty}(2m/(2m+1)) = 1$, we have

$$\frac{\pi}{2} = \lim_{m\to\infty} \frac{2^2 4^2 \ldots (2m-2)^2 2m}{3^2 5^2 \ldots (2m-1)^2}.$$

Hence

$$\sqrt{\frac{\pi}{2}} = \lim_{m\to\infty} \frac{2.4. \ldots .(2m-2)\sqrt{2m}}{3.5. \ldots .(2m-1)} = \lim_{m\to\infty} \frac{2^2 4^2 \ldots (2m-2)^2 2m\sqrt{2m}}{(2m)!}$$

$$= \lim_{m\to\infty} \frac{(2.4. \ldots .2m)^2}{(2m)!\sqrt{2m}} = \lim_{m\to\infty} \frac{2^{2m}(m!)^2}{(2m)!\sqrt{2m}},$$

and so we conclude that

$$\sqrt{\pi} = \lim_{m\to\infty} \frac{2^{2m}(m!)^2}{(2m)!\sqrt{m}},$$

exactly as required.

Wallis's formula will play a crucial role in the next section.

9.2 Stirling's Formula

We have already made use of Stirling's extraordinary asymptotic formula

$$n! \sim \sqrt{2\pi n} \left(\frac{n}{e} \right)^n. \tag{9.4}$$

In fact, even for modest values of n the approximation is quite good: for $n = 10$ the error is only 0.8%, and for $n = 100$ the error drops to 0.08%. Calculations suggest that the formula is always an underestimate for $n!$, and this will be confirmed by our proof. Exercise 6.19 goes some way to making the formula seem plausible, and indeed our approach depends on a more careful consideration of $\int_1^n \log x\, dx$.

Let us therefore begin by examining the function $x \mapsto \log x$ in the interval $[k, k+1]$ between two positive integers.

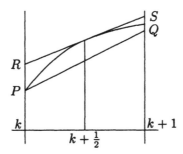

The concavity of the function is downwards, since the second derivative is negative, and so the area a_k under the curve is greater than the trapezial area s_k under the secant PQ. That is, by the standard formula for the area of a trapezium,

$$a_k > s_k = \frac{1}{2}[\log k + \log(k+1)]. \tag{9.5}$$

On the other hand, if RS is the tangent to the curve at $x = k + \frac{1}{2}$, the area a_k is less than the trapezial area t_k under RS. The tangent has gradient $1/(k+\frac{1}{2})$ (the value of the derivative at $x = k + \frac{1}{2}$) and so has equation

$$y - \log(k + \tfrac{1}{2}) = \frac{2}{2k+1}(x - k - \tfrac{1}{2}).$$

It meets $x = k$ in $R\big(k, \log(k + \frac{1}{2}) - 1/(2k + 1)\big)$, and meets $x = k + 1$ in $S\big(k + 1, \log(k + \frac{1}{2}) + 1/(2k + 1)\big)$. It follows that

$$t_k = \log(k + \tfrac{1}{2}) > a_k. \tag{9.6}$$

Now let $d_k = a_k - s_k$, the difference between the area under the curve and the area of the lower trapezium. Then

$$d_k < t_k - s_k = \log(k + \tfrac{1}{2}) - \frac{1}{2}\log k - \frac{1}{2}\log(k+1)$$

$$= \frac{1}{2}[\log(k + \tfrac{1}{2}) - \log k] - \frac{1}{2}[\log(k+1) - \log(k + \tfrac{1}{2})]$$

$$= \frac{1}{2}\log\left(1 + \frac{1}{2k}\right) - \frac{1}{2}\log\left(1 + \frac{1}{2(k + \frac{1}{2})}\right)$$

$$< \frac{1}{2}\log\left(1 + \frac{1}{2k}\right) - \frac{1}{2}\log\left(1 + \frac{1}{2(k+1)}\right). \tag{9.7}$$

Now let

$$A_n = \sum_{k=1}^{n-1} a_k = \int_1^n \log x \, dx = n \log n - n + 1 , \tag{9.8}$$

$$S_n = \sum_{k=1}^{n-1} s_k = \frac{1}{2}[(\log 1 + \log 2) + (\log 2 + \log 3) + \cdots + (\log(n-1) + \log n)]$$

$$= \log 2 + \log 3 + \cdots + \log(n-1) + \frac{1}{2} \log n$$

$$= \log(n!) - \frac{1}{2} \log n , \tag{9.9}$$

and let

$$D_n = A_n - S_n = \sum_{k=1}^{n-1} (a_k - s_k) .$$

Then certainly (D_n) is an increasing sequence. It is also bounded above, for, by (9.6) and (9.7),

$$D_n < \sum_{k=1}^{n-1} (t_k - s_k)$$

$$< \frac{1}{2} \left[\left(\log \frac{3}{2} - \log \frac{5}{4} \right) + \left(\log \frac{5}{4} - \log \frac{7}{6} \right) + \cdots \right.$$

$$\left. \cdots + \left(\log \left(1 + \frac{1}{2(n-1)} \right) - \log \left(1 + \frac{1}{2n} \right) \right) \right]$$

$$= \frac{1}{2} \left(\log \frac{3}{2} - \log \left(1 + \frac{1}{2n} \right) \right) ,$$

since all other terms cancel.

Hence

$$D_n < \frac{1}{2} \log \frac{3}{2}$$

for all n, and it follows that (D_n) has a limit. Let us denote this limit by D. Moreover,

$$D - D_n < \frac{1}{2} \log \left(1 + \frac{1}{2n} \right) , \tag{9.10}$$

since

$$D - D_n = \sum_{k=n}^{\infty} (a_k - s_k) < \sum_{k=n}^{\infty} (t_k - s_k)$$

$$< \frac{1}{2} \left[\log \left(1 + \frac{1}{2n} \right) - \log \left(1 + \frac{1}{2(n+1)} \right) \right]$$

$$+ \frac{1}{2} \left[\log \left(1 + \frac{1}{2(n+1)} \right) - \log \left(1 + \frac{1}{2(n+2)} \right) \right] + \cdots$$

$$= \frac{1}{2} \log \left(1 + \frac{1}{2n} \right) . \tag{9.11}$$

Now, by (9.8) and (9.9),

$$D_n = A_n - S_n = n \log n - n + 1 - \log(n!) + \frac{1}{2} \log n,$$

and so

$$\log(n!) = (n + \tfrac{1}{2}) \log n - n + (1 - D_n). \tag{9.12}$$

Write $\delta_n = e^{1-D_n}$; then (δ_n) is decreasing, with limit $\delta = e^{1-D}$. From (9.12) it follows that

$$n! = \delta_n n^{n+\frac{1}{2}} e^{-n}. \tag{9.13}$$

From (9.11), and from the obvious result that $(1 + 1/4n)^2 > 1 + 1/2n$, we deduce that

$$1 < \frac{\delta_n}{\delta} = e^{D-D_n} < e^{\frac{1}{2} \log(1+1/2n)} = \sqrt{(1 + 1/2n)} < 1 + \frac{1}{4n}.$$

Hence $\delta < \delta_n < \delta(1 + 1/4n)$, and so, from (9.13),

$$\delta n^{n+\frac{1}{2}} e^{-n} < n! < \delta n^{n+\frac{1}{2}} e^{-n} \left(1 + \frac{1}{4n}\right). \tag{9.14}$$

It remains to find the value of δ. If we use (9.13) to substitute for $n!$ in Wallis's formula (9.1), we obtain

$$\sqrt{\pi} = \lim_{n\to\infty} \frac{(n!)^2 2^{2n}}{(2n)! \sqrt{n}}$$

$$= \lim_{n\to\infty} \frac{\delta_n^2 n^{2n+1} e^{-2n} 2^{2n}}{\delta_{2n}(2n)^{2n+\frac{1}{2}} e^{-2n} \sqrt{n}}$$

$$= \lim_{n\to\infty} \frac{\delta_n^2}{\delta_{2n} \sqrt{2}} = \frac{\delta}{\sqrt{2}},$$

and so $\delta = \sqrt{2\pi}$. From (9.14) we now obtain a refined version of Stirling's formula, in the form of an inequality:

$$\sqrt{2\pi n} \left(\frac{n}{e}\right)^n < n! < \sqrt{2\pi n} \left(\frac{n}{e}\right)^n \left(1 + \frac{1}{4n}\right).$$

From this the asymptotic formula (9.4) follows immediately.

EXERCISES

9.1 Determine

$$\lim_{n\to\infty} \frac{1}{n} (n!)^{1/n}.$$

9.3 A Continuous, Nowhere Differentiable Function

Our final example is of what is sometimes called a "pathological" function. We show that there exists a function $f : [0, 1] \to \mathbb{R}$ which at every x in $(0, 1)$ is continuous but not differentiable. Weierstrass gave the first example of such a function; the one we now describe is much later, and is due to van der Waerden.

For each real number x let $\langle x \rangle$ denote the difference between x and the integer nearest to x: that is,

$$\langle x \rangle = \min \left\{ x - \lfloor x \rfloor, \lceil x \rceil - x \right\}.$$

(See page 66 for the definitions of $\lfloor x \rfloor$ and $\lceil x \rceil$.) The function $x \mapsto \langle x \rangle$ is continuous, with graph

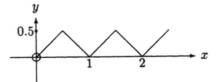

For each n in $\mathbb{N}$, let

$$f_n(x) = \sum_{r=1}^{n} \frac{\langle 10^r x \rangle}{10^r} \quad (x \in [0, 1]).$$

Then f_n is continuous. We now show that the sequence (f_n) converges uniformly in $[0, 1]$. We do this by demonstrating that (f_n) is a Cauchy sequence. Let $m > n$. Then, for all x in $[0, 1]$,

$$|f_m(x) - f_n(x)| = \left| \sum_{r=n+1}^{m} \frac{\langle 10^r x \rangle}{10^r} \right| \leq \sum_{r=n+1}^{m} \frac{|\langle 10^r x \rangle|}{10^r}$$

$$< \sum_{r=n+1}^{m} \frac{1}{10^r} < \frac{1}{10^{n+1}} \cdot \frac{1}{1 - (1/10)} = \frac{1}{9.10^n},$$

and it follows that for every $\epsilon > 0$ there exists N with the property that $\|f_m - f_n\| < \epsilon$ for all $m > n > N$. It follows that the sequence (f_n) tends to a function f, and, by Theorem 7.5, f too is continuous at all points in $[0, 1]$.

Suppose now that $f'(x)$ exists for some x in $(0, 1)$. Then, for every sequence (h_n) with limit 0, the sequence (b_n), where

$$b_n = \frac{f(x + h_n) - f(x)}{h_n},$$

has limit $f'(x)$. We derive a contradiction by choosing a sequence (h_n) for which this is not so. Suppose that x has decimal expansion $0.a_1 a_2 a_3 \ldots$, and define

$$h_n = \begin{cases} -1/10^n & \text{if } a_n = 4 \text{ or } a_n = 9 \\ 1/10^n & \text{otherwise.} \end{cases}$$

If $r \geq n$, then $10^r(x + h_n)$ has the same fractional part as $10^r x$, and so

$$\langle 10^r(x + h_n) \rangle = \langle 10^r x \rangle \quad (r \geq n). \tag{9.15}$$

If $r < n$, then the fractional part of $10^r(x + h_n)$ differs from that of $10^r x$ in (and only in) the $(n - r)$th place. Denote the $(n - r)$th entry of $10^r x$ by p. If $p = 4$ or 9, then the entry for $10^r(x + h_n)$ is $p - 1$; otherwise it is $p + 1$. In any event

$$\langle 10^r(x + h_n) \rangle - \langle 10^r x \rangle = \pm \frac{1}{10^{n-r}}.$$

Now, for all $m > n$,

$$\begin{aligned} \frac{f_m(x + h_n) - f_m(x)}{h_n} &= \sum_{r=1}^{m} \frac{\langle 10^r(x + h_n) \rangle - \langle 10^r x \rangle}{10^r h_n} \\ &= \sum_{r=1}^{n-1} \frac{\langle 10^r(x + h_n) \rangle - \langle 10^r x \rangle}{10^r h_n} \quad \text{(by virtue of (9.15))} \\ &= \sum_{r=1}^{n-1} \pm \frac{1/10^{n-r}}{10^r(1/10^n)} = \sum_{r=1}^{n-1} (\pm 1). \end{aligned}$$

Letting m tend to infinity, we see that $b_n = \sum_{r=1}^{n-1}(\pm 1)$. Certainly (b_n) is not a Cauchy sequence, and so cannot have limit $f'(x)$. We are forced to conclude that f is not differentiable at x.

Solutions to Exercises

Chapter 1

1.1 If $x = m/n$ were rational, then $x^2 = m^2/n^2$ would also be rational. The square of $\sqrt{2}+\sqrt{3}$ is $5+2\sqrt{6}$. If $5+2\sqrt{6} = p/q \in \mathbb{Q}$, then $\sqrt{6} = (p-5q)/10q \in \mathbb{Q}$, and this is not the case.

1.2 $A \cap B = \{1,3\}$, $A \cup B = \{1,2,3,4,5,7\}$, $X \setminus A = \{2,4,6,8\}$, $X \setminus B = \{5,6,7,8\}$. $X \setminus (A \cap B) = (X \setminus A) \cup (X \setminus B) = \{2,4,5,6,7,8\}$, $X \setminus (A \cup B) = (X \setminus A) \cap (X \setminus B) = \{6,8\}$.

1.3 $437 = 19 \times 23$, $493 = 17 \times 29$.

1.4 a) True. If $x = m/n \in \mathbb{Q}$ and $x+y = p/q \in \mathbb{Q}$, then $y = (pn-qm)/qn \in \mathbb{Q}$, a contradiction.

  b) False. Take $x = 0$. If we had insisted that $x \neq 0$ this would have been true.

  c) False. Take $x = \sqrt{2}$, $y = 1 - \sqrt{2}$.

  d) False. Take $x = y = \sqrt{2}$.

  e) False. Take $x = \sqrt{2}$, $y = \sqrt{3}$.

  f) False. Again take $x = \sqrt{2}$, $y = \sqrt{3}$.

1.5 Take $u = x + (1/\sqrt{2})(y - x)$.

1.6 Let N be a positive integer such that $N > 1/(y - x)$. Then the difference between successive members of the sequence

$$\ldots, -3/N, -2/N, -1/N, 0, 1/N, 2/N, \ldots$$

is less than $y - x$, and so $x < q < y$ for at least one rational number $q = M/N$.

1.7 True for $n = 1$. Let $n \geq 1$ and suppose true for n. Then $1^2 + \cdots + (n+1)^2 = (1/6)n(n+1)(2n+1) + (n+1)^2 = (1/6)(n+1)(2n^2 + n + 6n + 6) = (1/6)(n+1)[(n+1)+1][2(n+1)+1]$.

1.8 True for $n = 1$. Let $n \geq 1$ and suppose true for n. Then $1^3 + \cdots + (n+1)^3 = (1/4)n^2(n+1)^2 + (n+1)^3 = (1/4)(n+1)^2(n^2 + 4n + 4) = (1/4)(n+1)^2[(n+1)+1]^2$.

1.9 True for $n = 1$. Let $n \geq 1$ and suppose true for n. Then $a + (a+d) + \cdots + (a+nd) = na + \frac{1}{2}n(n-1)d + a + nd = (n+1)a + \frac{1}{2}[n(n-1) + 2n]d = (n+1)a + \frac{1}{2}(n+1)[(n+1) - 1]d$.

1.10 True for $n = 1$. Let $n \geq 1$ and suppose true for n. Then $a + ar + \cdots + ar^n = [a(1-r^n)]/(1-r) + ar^n = [a - ar^n + ar^n - ar^{n+1}]/(1-r) = [a(1-r^{n+1})]/(1-r)$.

1.11 If $n = 1$ then the right hand side is $(x^2 - 2x + 1)/(x-1)^2 = 1$. So true for $n = 1$. Let $n \geq 1$ and suppose true for n. Then

$$1 + 2x + \cdots + (n+1)x^n = \frac{nx^{n+1} - (n+1)x^n + 1}{(x-1)^2} + (n+1)x^n$$

$$= \frac{nx^{n+1} - (n+1)x^n + 1 + (n+1)x^{n+2} - 2(n+1)x^{n+1} + (n+1)x^n}{(x-1)^2}$$

$$= \frac{(n+1)x^{n+2} - (n+2)x^{n+1} + 1}{(x-1)^2}.$$

1.12 $(n+1)^4 < 4n^4$ if and only if $(n+1)/n < \sqrt{2}$, that is, if and only if $n(\sqrt{2} - 1) > 1$, that is, if and only if n (being an integer) is at least 3. $4^n > n^4$ if $n = 5$. Let $n \geq 5$, and suppose true for n. Then $4^{n+1} = 4.4^n > 4n^4 > (n+1)^4$.

1.13 The formula is true for $n = 1$. Let $n \geq 1$ and suppose true for n. Then $q_{n+1} = 3q_n - 1 = (3/2)(3^n + 1) - 1 = (1/2)(3^{n+1} + 3 - 2) = (1/2)(3^{n+1} + 1)$.

1.14 The formula is true for $n = 0$ and $n = 1$. Suppose that it is true for all $k < n$. Then $a_n = 4[2^{n-2}(n+1) - 2^{n-3}n] = 2^{n-1}[2n + 2 - n] = 2^{n-1}(n+2)$.

1.15 a) $x^2 + 4x + 5 = (x+2)^2 + 1 \geq 1 > 0$. b) $x^2 + 5xy + 7y^2 = (x + (5/2)y)^2 + (3/4)y^2 \geq 0$. c) $a^2 + b^2 + c^2 + (1/a^2) + (1/b^2) + (1/c^2) - 6 = [a - (1/a)]^2 + [a - (1/b)]^2 + [a - (1/c)]^2 \geq 0$. Equality occurs if and only if $a = 1/a$, $b = 1/b$, $c = 1/c$, that is, if and only if $a, b, c \in \{-1, 1\}$.

1.16 $2a \leq a + b \leq 2b$, and so $2ab/2b \leq 2ab/(a+b) \leq 2ab/2a$; that is, $a \leq H \leq b$. Also, $G^2 - H^2 = ab - \left(4a^2b^2/(a+b)^2\right) = [ab(a-b)^2]/(a+b)^2 \geq 0$.

1.17 Trick question! $A = \{(x,y) : (x+1)^2 + (y-2)^2 + 1 = 0\} = \emptyset$, $B = \{(x,y) : (x+2y+1)^2 + y^2 + 2 = 0\} = \emptyset$.

1.18 $|x - a| < \delta \iff [x - a < \delta \text{ and } -(x-a) < \delta] \iff a - \delta < x < a + \delta$.

1.19 From Exercise 1.10, $a + ar + \cdots ar^n = a/(1-r) - ar^{n+1}/(1-r) < a/(1-r)$.

1.20 Put $a = 1$ and $r = |x_2/x_1|$ in the previous inequality, to obtain $1 + |x_2/x_1| + \cdots |x_2/x_1|^{n-1} < 1/(1 - |x_2/x_1|)$. The left hand side is not greater than n times its smallest term, so not greater than $n|x_2/x_1|^{n-1}$. Hence $n|x_2/x_1|^{n-1} < 1/(1 - |x_2/x_1|) = |x_1|/(|x_1| - |x_2|)$.

1.21 $(3x+2)/(x+1) < 1 \iff (3x+2)/(x+1) - 1 < 0 \iff (3x + 2 - x - 1)/(x+1) < 0 \iff (2x+1)/(x+1) < 0 \iff x \in (-1, -1/2]$.

1.22 If $x \le y$, then $|x - y| = y - x$, so $(1/2)(x + y + |x - y|) = y = \max\{x,y\}$, $(1/2)(x + y - |x - y|) = x = \min\{x,y\}$. The case $x \ge y$ is similar.

1.23 $|ab - cd| = |b(a - c) + c(b - d)| \le |b|\,|a - c| + |c|\,|b - d|$. So with the given information $|ab - cd| \le |b|\,|a - c| + |c|\,|b - d| < K(\epsilon/2K) + L(\epsilon/2L) = \epsilon$.

1.24 Since $|ab| > k^2$, $|(1/a) - (1/b)| = |a - b|/|ab| \le |a - b|/k^2$.

Chapter 2

2.1 The following table gives the answer:

ϵ	0.01	0.001	0.0001
N	10,000	1,000,000	100,000,000

2.2 If $k = 0$ then the sequence is $(1,1,1,\ldots)$, with limit 1. If $k > 0$, then $|1/n^k - 0| < \epsilon$ whenever $n > (1/\epsilon)^{1/k}$. Hence $(1/n^k) \to 0$. If $k = -l < 0$, then $1/n^k = n^l > M$ for any given $M > 0$, provided $n > M^{1/l}$. (Note that, for any positive a in $\mathbb{R}$ and any positive integer k, we can define $a^{1/k}$ as $\sup\{x \in \mathbb{R} : x^k < a\}$.)

2.3 $n/(n^2 + 1) < 0.0001$ if and only if $n^2 - 10,000n + 1 > 0$, that is, if and only if $n > 5,000 + \sqrt{25,000,000 - 1}$, that is, if and only if $n > 9,999$.

2.4 $n^2 + 2n \ge 9999 \iff n \ge -1 + \sqrt{1 + 9999} = 99$.

2.5 $a_n \to -\infty$ if and only if, for every $K > 0$ there exists a positive integer N with the property that $a_n < -K$ for every $n > N$.

2.6 The formula is correct for $n = 1$ and $n = 2$. Suppose that it is true for all $k < n$. Then $a_n = (1/2)[2 + 4(-(1/2))^{n-1} + 2 + 4(-(1/2))^{n-2}] =$

$(1/2)\big[4 - 8(-(1/2))^n + 16(-(1/2))^n\big] = 2 + 4(-(1/2))^n$. It is then clear that $(a_n) \to 2$.

2.7 Suppose that $\beta > B$. There exists N such that $|b_n - \beta| < \beta - B$ for all $n > N$. Thus $\beta - b_n < \beta - B$ and so $b_n > B$ for all $n > N$. This is a contradiction to the definition of B.

2.8 If $|a_n| \le A$, then $-A \le a_n \le A$. So "bounded" implies "bounded above and below". Conversely, suppose that $A \le a_n \le B$. If $0 \le A \le B$ then $|a_n| = a_n \le B$. If $A \le B \le 0$ then $|a_n| = -a_n \le -A$. If $A \le 0 \le B$ then $|a_n| \le \max\{|A|, |B|\}$. So in all cases (a_n) is bounded.

2.9 We know that for every $\epsilon > 0$ there exists a positive integer N such that $|a_n - \alpha| < \epsilon$ for every $n > N$. It follows that there exists a positive integer N', namely $N' = N - 1$, such that $|b_n - \alpha| < \epsilon$ for every $n > N'$. Thus $(b_n) \to \alpha$.

2.10 Since $(a_n) \to \alpha$, for every $\epsilon > 0$ there exists N such that $|a_n - \alpha| < \epsilon$ for all $n > N$. There exists an integer M with the property that $b_M = a_K$, with $K \ge N$. Then, for all $m > M$, $b_m = a_k$ for some $k > N$, and so $|b_m - \alpha| < \epsilon$.

2.11 Suppose, for a contradiction, that $L < 0$. Taking $\epsilon = |L|/2 = -L/2$, we know that there exists a positive integer N such that $|a_n - L| < |L|/2$ for all $n > N$. This implies in particular that $a_n < L/2 < 0$ —a contradiction.

2.12 Suppose first that $L > 0$. Since a_n is positive for all n, $|\sqrt{a_n} - \sqrt{L}| = (|a_n - L|)/(\sqrt{a_n} + \sqrt{L}) < (1/\sqrt{L})|a_n - L|$. Choosing N so that $|a_n - L| < (\sqrt{L})\epsilon$ for all $n > N$, we see that $|\sqrt{a_n} - \sqrt{L}| < \epsilon$ for all $n > N$. So $(\sqrt{a_n}) \to \sqrt{L}$.

Suppose now that $L = 0$. If, for a given ϵ, we choose N so that $|a_n| < \epsilon^2$ for every $n > N$, then $|\sqrt{a_n}| < \epsilon$, and so $(\sqrt{a_n}) \to 0$.

2.13 Since $(a_n - x_n)$ is a positive sequence with limit $\xi - \alpha$, we deduce from Exercise 2.10 that $\xi - \alpha > 0$. Similarly $\beta - \xi \ge 0$.

2.14 By Theorems 2.1 and 2.8,

$$\lim_{n \to \infty} \max\{a_n, b_n\} = \max\{\alpha, \beta\}, \quad \lim_{n \to \infty} \min\{a_n, b_n\} = \min\{\alpha, \beta\}.$$

Let $a_n = (-1)^n$, $b_n = (-1)^{n+1}$. Then $\max\{a_n, b_n\} = 1$, $\min\{a_n, b_n\} = -1$. So (a_n) and (b_n) both diverge, while $(\max\{a_n, b_n\})$ and $(\min\{a_n, b_n\})$ are both convergent. The final statement follows from the observation that

$$b_n = \max\{a_n, b_n\} + \min\{a_n, b_n\} - a_n.$$

2.15 $a^{1/n} = 1 + h_n$, where $h_n > 0$. By the binomial theorem, $a = (1+h_n)^n > 1 +$ nh_n (since all the remaining terms are positive). Thus $0 < h_n < (a-1)/n$, and so, by the sandwich principle, $(h_n) \to 0$. Hence $(a^{1/n}) \to 1$. If $a = 1$ the sequence is constant, with limit 1. If $0 < a < 1$ then $a = 1/b$, where $b > 1$. Since $(b^{1/n}) \to 1$, it follows that $a^{1/n} = 1/b^{1/n} \to 1$ as $n \to \infty$.

2.16 Certainly $(2^n + 3^n)^{1/n} > (3^n)^{1/n} = 3$ for all $n \geq 1$. On the other hand, $(2/3)^n < 1$ for all $n \geq 1$, and so $(2^n + 3^n)^{1/n} = 3[1 + (2/3)^n]^{1/n} < 3(2^{1/n})$. Thus $3 < (2^n + 3^n)^{1/n} < 3(2^{1/n})$, and so, by the sandwich principle and the result of the previous exercise, $((2^n + 3^n)^{1/n}) \to 3$.

2.17 $|a_n - \alpha| = |a_n^3 - \alpha^3|/(a_n^2 + a_n\alpha + \alpha^2) = |a_n^3 - \alpha^3|/[(a_n + \frac{1}{2}\alpha)^2 + \frac{3}{4}\alpha^2] \leq$ $|a_n^3 - \alpha^3|/(3/4)\alpha^2$. So, for a given $\epsilon > 0$, choose N so that, for all $n > N$, $|a_n^3 - \alpha^3| < (3/4)\alpha^2\epsilon$. Then $|a_n - \alpha| < \epsilon$, and so $(a_n) \to \alpha$.

a) Yes. For a given ϵ, choose N so that $|a_n^3| < \epsilon^3$ for all $n > N$. Then $|a_n| < \epsilon$.

b) No. Consider $a_n = (-1)^n$.

2.18 a) It is clear that $a_n \geq 0$ for all n. Certainly $a_1 < 1$. Suppose inductively that $a_n < 1$. Then $1 - a_{n+1} = 1 - (3a_n + 1)/(a_n + 3) = 2(1 - a_n)/(a_n + 3) > 0$.

b) $a_{n+1} - a_n = (3a_n + 1)/(a_n + 3) - a_n = (1 - a_n^2)/(a_n + 3) > 0$.

c) Since the sequence is increasing, and bounded above by 1, it has a limit α, which must be non-negative, and satisfies $\alpha = (3\alpha + 1)/(\alpha + 3)$. Thus $\alpha = 1$.

2.19 a) It is clear that $a_1^2 > 3$. Suppose inductively that $a_n^2 > 3$. Then $a_{n+1}^2 - 3 = \cdots = (a_n^2 - 3)/(a_n + 2)^2 > 0$.

b) $a_n - a_{n+1} = a_n - (2a_n + 3)/(a_n + 2) = (a_n^2 - 3)/(a_n + 2) > 0$.

c) Since (a_n) is decreasing and bounded below by $\sqrt{3}$, it has a positive limit α, satisfying $\alpha = (2\alpha + 3)/(\alpha + 2)$. Thus $\alpha = \sqrt{3}$.

2.20 If a limit α exists, then it is positive, and satisfies $\alpha^2 = 2 + 2\alpha$; thus $\alpha = 1 + \sqrt{3}$. The sequence begins (approximately) $(2, 2.45, 2.62, \ldots)$, which suggests that it might be monotonic increasing. Were that to be the case, $1 + \sqrt{3}$ would be the supremum. It is clear that $a_1 < 1 + \sqrt{3}$, so suppose inductively that $0 < a_n < 1 + \sqrt{3}$. Then $a_{n+1}^2 = 2 + 2a_n < 2 + 2(1 + \sqrt{3}) = 4 + 2\sqrt{3} = (1 + \sqrt{3})^2$, and so $a_{n+1} < 1 + \sqrt{3}$. Thus, since it is clear that every a_n is positive, $0 < a_n < 1 + \sqrt{3}$ for all n. Since $x^2 - 2x - 2$ takes negative values in the interval $(1 - \sqrt{3}, 1 + \sqrt{3})$, $a_n^2 - a_{n+1}^2 = a_n^2 - 2a_n - 2 < 0$, and so (a_n) is monotonic increasing. Hence $\lim_{n \to \infty} a_n$ exists, and equals $1 + \sqrt{3}$.

2.21 By the arithmetic-geometric inequality (1.26), we know that $a_n \geq b_n$ for all $n \geq 2$. Since $a_{n+1} - a_n = \frac{1}{2}(a_n + b_n) - a_n = \frac{1}{2}(b_n - a_n) \leq 0$, it follows that (a_n) is decreasing. Since $b_{n+1}/b_n = \sqrt{a_n b_n}/b_n = \sqrt{a_n/b_n} \geq 1$, it follows that (b_n) is increasing. Since $a_n \geq b_n \geq b_2$ and $b_n \leq a_n \leq a_2$, both sequences converge: say $(a_n) \to \alpha$, $(b_n) \to \beta$. But then $\alpha = (\alpha + \beta)/2$, and so $\alpha = \beta$.

2.22 Let $m > n$. For the first sequence, $|a_m - a_n| = |(-1)^m/m - (-1)^n/n| \leq (1/m) + (1/n) \leq 2/n$. So if, for a given $\epsilon > 0$, we choose $N_1 \geq 2/\epsilon$, then $|a_m - a_n| < \epsilon$ for all $m > n > N_1$. In the same way, for the second sequence $|a_m - a_n| \leq 2/2^{n-\frac{1}{2}} \leq 1/2^{n-2} < \epsilon$ for all $m > n > N_2$, where N_2 is chosen so that $2^{N_2-2} > 1/\epsilon$.

2.23 Let m, n and N be positive integers such that $m > n > N$. The given property implies that $a_2 \neq a_1$. Then $|a_m - a_n| = |(a_{n+1} - a_n) + (a_{n+2} - a_{n+1}) + \cdots + (a_m - a_{m-1})| \leq |a_{n+1} - a_n| + |a_{n+2} - a_{n+1}| + \cdots + |a_m - a_{m-1}| \leq |a_2 - a_1|(k^{n-1} + k^n + \cdots + k^{m-2}) < k^{n-1}|a_2 - a_1|/(1 - k) < k^{N-1}|a_2 - a_1|/(1 - k)$. If, for a given $\epsilon > 0$, we now choose N so that $K^{N-1} < \epsilon(1 - k)/|a_2 - a_1|$, we find that $|a_m - a_n| < \epsilon$ for all $m > n > N$.

To see that $k = 1$ is not good enough, let $a_n = \sqrt{n}$. Then $|a_{n+1} - a_n| = 1/(\sqrt{n+1} + \sqrt{n}) < 1/(\sqrt{n} + \sqrt{n-1}) = |a_n - a_{n-1}|$, but (a_n) is not a Cauchy sequence.

2.24 Consider an arbitrary but fixed $n > N$, and let $b_m = |a_m - a_n|$ $(m \geq n)$. By assumption, $b_m < \epsilon$ for all m, and so $\lim_{m \to \infty} b_m \leq \epsilon$. That is, $|\alpha - a_n| \leq \epsilon$.

2.25 a) $b_n b_{n-1} = (g_{n+1}/g_n)(g_n/g_{n-1}) = g_{n+1}/g_{n-1} = (g_n + 2g_{n-1})/g_{n-1} = b_{n-1} + 2$.

   b) Divide by b_{n-1} to obtain $b_n = 1 + (2/b_{n-1})$.

   c) Observe first that $g_n > g_{n-1}$, and so $b_{n-1} > 1$. Thus it follows from part a) that $b_n b_{n-1} > 3$. Hence $|b_{n+1} - b_n| = |1 + (2/b_n) - 1 - (2/b_{n-1})| = 2|(b_n - b_{n-1})/(b_n b_{n-1})| < (2/3)|b_n - b_{n-1}|$.

   d) It follows from Exercise 2.23 that (b_n) is a Cauchy sequence, and so has a limit β. From part a) it follows that $\beta^2 = \beta + 2$, and so $\beta = 2$.

2.26 $L_{n+1} - L_n = 1/(2n + 1) + 1/(2n + 2) - 1/(n + 1) \geq 1/(2n + 2) + 1/(2n + 2) - 1/(n + 1) = 0$, so (L_n) is monotonic increasing. Also $1/(2n) + 1/(2n) + \cdots + 1/(2n) < L_n < 1/n + 1/n + \cdots + 1/n$, and so $1/2 < L_n < 1$ for all n. Hence $(L_n) \to \alpha$, where $1/2 \leq \alpha \leq 1$.

2.27 For all n, $1/n(n+2) = (1/2)[(1/n) - (1/(n+2))]$. Hence $\sum_{n=1}^{N}(1/n(n+2)) = (1/2)[(1 - (1/3)) + ((1/2) - (1/4)) + \cdots + ((1/N) - (1/(N + 2)))]$. Most terms cancel, and so $\sum_{n=1}^{N}(1/n(n+2)) = (1/2)[1 + (1/2) - 1/(N+1) - 1/(N+2)] \to \frac{3}{4}$ as $N \to \infty$.

2.28 $1/n! - 1/(n+1)! = [(n+1) - 1]/(n+1)! = n/(n+1)!$, and so $\sum_{n=1}^{N}(n/(n+1)!) = \big((1 - (1/2!)) + ((1/2!) - (1/3!))\big) + \cdots + \big((1/N!) - (1/(N+1)!)\big) = 1 - \big(1/(N+1)!\big) \to 1$ as $N \to \infty$. Observing that

$$n/(n+1)! < (n+1)/(n+1)! = 1/n! < (n-1)/n!,$$

we see that $\sum_{n=0}^{\infty}(1/n!) = 1 + \sum_{n=1}^{\infty}(1/n!) \ge 1 + \sum_{n=1}^{\infty}(n/(n+1)!) = 2$, and $\sum_{n=0}^{\infty}(1/n!) = 1 + 1 + \sum_{n=2}^{\infty}(1/n!) \le 2 + \sum_{n=2}^{\infty}((n-1)/n!) = 2 + \sum_{n=1}^{\infty}(n/(n+1)!) = 3$.

2.29 Let $a_n = 1 + \frac{1}{2} + \cdots + \frac{1}{n}$. Then $|a_{n+1} - a_n| = 1/n(n+1) \to 0$ as $n \to \infty$, but (a_n) cannot be a Cauchy sequence, since it does not converge.

2.30 Let $\sum_{n=1}^{N} a_n = A_N$, $\sum_{n=1}^{N} b_n = B_N$. Then $\sum_{n=1}^{N}(a_n + b_n) = A_N + B_N$, and so $\sum_{n=1}^{\infty}(a_n + b_n) = \lim_{N \to \infty}(A_N + B_N) = \lim_{N \to \infty} A_N + \lim_{N \to \infty} B_N = A + B$. Also, $\sum_{n=1}^{N}(ka_n) = kA_N \to kA$ as $N \to \infty$.

2.31 a) False. The harmonic series is a counterexample.

   b) False. Take $a_n = 1/n$ and $b_n = 1/n^2$.

   c) True. It follows from Theorem 2.28.

   d) True. Since $(a_n) \to 0$, we may suppose that $0 \le a_n < 1$ beyond a certain point. Hence $a_n^2 < a_n$, and so $\sum_{n=1}^{\infty} a_n^2$ is convergent by the comparison test.

   e) False. Let $a_{2^k} = 1/k$ for $k = 0, 1, 2, \ldots$, and otherwise let $a_n = 0$. Then $a_{n+1} + \cdots + a_{2n} = 1/k$, where k is defined uniquely by the property $n+1 \le 2^k \le 2n$. Since $k \to \infty$ as $n \to \infty$, we have that $\lim_{n \to \infty}(a_{n+1} + \cdots + a_{2n}) = 0$. On the other hand, $\sum_{n=1}^{\infty} a_n = \sum_{k=1}^{\infty}(1/k)$ is divergent.

2.32 Let $\sum_{n=1}^{N} a_n = A_N$, $\sum_{n=1}^{N} b_n = B_N$. Then $A_N \le B_N$, and so $A = \lim_{N \to \infty} A_N \le \lim_{N \to \infty} B_N = B$.

2.33 Since $\sqrt{n+1}/(n^2 + 2) \sim 1/n^{3/2}$, the first series converges. So does the second series, since $(n+3)/(n^4 - 1) \sim 1/n^3$. The third series diverges, since $(n+1)/\sqrt{n^3 + 2} \sim 1/n^{1/2}$.

2.34 $\sqrt{n+1} - \sqrt{n} = 1/(\sqrt{n+1} + \sqrt{n}) \asymp 1/\sqrt{n}$, so the series is divergent. Alternatively, observe that $\sum_{n=1}^{N}(\sqrt{n+1} - \sqrt{n}) = \sqrt{N+1} - 1 \to \infty$ as $N \to \infty$. $\sum_{n=1}^{N}[(1/\sqrt{n}) - (1/\sqrt{n+1}] = 1 - (1/\sqrt{N+1}) \to 1$ as $N \to \infty$, so the series is convergent.

2.35 For the first series $a_{n+1}/a_n = [(n+1)!n^n]/[(n+1)^{n+1}n!] = (n/(n+1))^n = 1/[1 + (1/n)]^n \to 1/e$ as $n \to \infty$. (See Example 2.13.) Since $1/e < 1$, the series converges. For the second series $a_{n+1}/a_n = [(n+1)^3 n!]/[(n+1)! n^3] = [(n+1)/n]^3 [1/(n+1)] \to 0$ as $n \to \infty$, and so the series is convergent.

2.36 Certainly $\sum_{n=1}^{\infty}(a_n + b_n)$ is convergent. Since $\max\{a_n, b_n\} < a_n + b_n$, it follows from the comparison test that $\sum_{n=0}^{\infty} \max\{a_n, b_n\}$ is convergent.

2.37 This follows from the previous exercise, since $(a_n b_n)^{1/2} \le \max\{a_n, b_n\}$. To show the converse false, let

$$a_n = \begin{cases} 1 & \text{if } n \text{ is even} \\ 0 & \text{if } n \text{ is odd} \end{cases} \qquad b_n = \begin{cases} 1 & \text{if } n \text{ is odd} \\ 0 & \text{if } n \text{ is even.} \end{cases}$$

Neither $\sum_{n=0}^{\infty} a_n$ nor $\sum_{n=0}^{\infty} b_n$ converges, but $(a_n b_n)^{1/2} = 0$ for every n.

2.38 Let

$$a_n = \begin{cases} 1/n & \text{if } n \text{ is even} \\ 1/n^2 & \text{if } n \text{ is odd} \end{cases} \qquad b_n = \begin{cases} 1/n & \text{if } n \text{ is odd} \\ 1/n^2 & \text{if } n \text{ is even.} \end{cases}$$

Neither $\sum_{n=1}^{\infty} a_n$ nor $\sum_{n=0}^{\infty} b_n$ is convergent, but $\min\{a_n, b_n\} = 1/n^2$.

2.39 Suppose that $\lim_{n\to\infty}(a_{n+1}/a_n) = L > 1$. Choosing $\epsilon > 0$ so that $R = L - \epsilon > 1$, we obtain an integer N with the property that $a_{n+1}/a_n > R$ for all $n > N$. It follows that $a_n > a_{N+1} R^{n-N-1}$ for all $n > N$, and so, from the $(N+1)$th term onwards, we have a comparison with the divergent geometric series $\sum_{n=N+1}^{\infty} a_{N+1} R^{n-N-1}$. Hence the series $\sum_{n=1}^{\infty} a_n$ is divergent.

2.40 $a_{n+1}/a_n = (n+1)^k a^{n+1}/n^k a^n = [(n+1)/n]^k a \to a$ as $n \to \infty$, and so the series is convergent. From Theorem 2.20 it follows that $\lim_{n\to\infty} n^k a^n = 0$.

2.41 The conditions of the Leibniz test are satisfied in both cases, and so the series are convergent. The second series is absolutely convergent, the first is not.

2.42 Let $a_n = 1/n$ if n is odd, and $a_n = 1/n^2$ if n is even. Then $(a_n) \to 0$ but is not monotonic decreasing. $\sum_{n=1}^{\infty}(-1)^{n-1} a_n$ is divergent, since the sum of the positive terms increases without limit, while the sum of the negative terms can never exceed the sum of the convergent series $\sum_{n=1}^{\infty}(1/(2n)^2)$.

2.43 a) $S_{2n} = 1 + (1/2) + \cdots + (1/2n) - 2[(1/2) + (1/4) + \cdots + (1/2n)] = H_{2n} - H_n$.

b) $T_{3n} = [1 + (1/3) + (1/5) + \cdots + (1/(4n-1))] - [(1/2) + (1/4) + \cdots + (1/2n)] = [H_{4n} - (1/2)H_{2n}] - (1/2)H_n = [H_{4n} - H_{2n}] + (1/2)[H_{2n} - H_n] = S_{4n} + (1/2)S_{2n}$. It follows that the rearranged series has sum $3S/2$.

2.44 Let U_n be the sum of the series to n terms. Then $U_{3n} = [1 + (1/3) + \cdots + 1/(2n-1)] - [(1/2) + (1/4) + \cdots + (1/4n)] = [H_{2n} - (1/2)H_n] - (1/2)H_{2n} = (1/2)(H_{2n} - H_n) = (1/2)S_{2n}$. It follows that the rearranged series has sum $S/2$.

Chapter 3

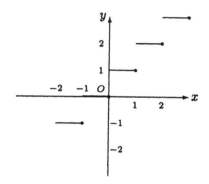

(The dots serve to indicate that the value of the function at each integer n is n.)

3.2 a)

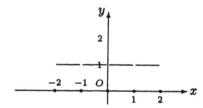

b)

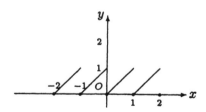

c)

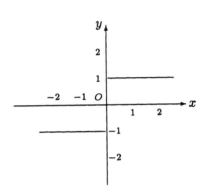

3.3 a) $\mathrm{dom}(g \circ f) = \{x \in \mathbb{R} : 2x - 3 \geq 0\} = [3/2, \infty)$, and $(g \circ f)(x) = \sqrt{2x - 3}$.

   b) $\mathrm{dom}(g \circ f) = \{x \in \mathbb{R} : \sqrt{4 - x^2} \neq 1\} = [-2, 2] \setminus \{\sqrt{3}, -\sqrt{3}\}$, and $(g \circ f)(x) = 1/[(4 - x^2)^{3/2} - 1]$.

   c) Since $f(x)$ is rational for every x, it follows that $\mathrm{dom}(g \circ f) = [0, \infty)$ and $(g \circ f)(x) = 1$ for every x in the domain.

3.4 a) $(f/g) \pm h = (f \pm g \cdot h)/g$.

   b) $(f/g) \cdot h = (f \cdot h)/g$.

   c) By repeated use of the above observations, we can delay the use of division to the very end, and a single division is enough.

3.5 $((f + g) \circ h)(x) = (f + g)(h(x)) = f(h(x)) + g(h(x)) = (f \circ h)(x) + (g \circ h)(x) = ((f \circ h) + (g \circ h))(x)$. Let $g(x) = h(x) = x$ and let $f(x) = x^2$. Then $(f \circ (g + h))(x) = (2x)^2 = 4x^2$, $((f \circ g) + (f \circ h))(x) = x^2 + x^2 = 2x^2$.

3.6 Let $q = f/g$, $r = h/k$, where f, g, h, k are polynomials. Then $q + r = (f \cdot k + g \cdot h)/(g \cdot k)$, $q \cdot r = (f \cdot h)/(g \cdot k)$, $q/r = (f \cdot k)/(g \cdot h)$ are all rational functions. If q and r are rational functions, then, for each x, $(q \circ r)(x)$ is obtained from $r(x)$ by repeated applications of addition, multiplication and division. The result is a rational expression in x.

3.7 For example,

$$(g \circ h)(x) = g(-1/x) = \frac{1}{-1/x} = -x, \quad (h \circ g)(x) = h(1/x) = \frac{1}{-1/x} = -x,$$

and so $g \circ h = h \circ g = f$.

3.8 Here are a few sample computations.

$$(f \circ g)(x) = f\left(1 - \frac{1}{x}\right) = 1 / \left(1 - \left(1 - \frac{1}{x}\right)\right) = 1 / \frac{1}{x} = x,$$

$$(g \circ f)(x) = g\left(\frac{1}{1-x}\right) = 1 - 1 \Big/ \frac{1}{1-x} = 1 - (1-x) = x;$$

thus $f \circ g = g \circ f = i$.

$$(f \circ p)(x) = f(1/x) = \frac{1}{1-(1/x)} = \frac{x}{x-1}, \quad (p \circ f)(x) = p(1/(1-x)) = 1-x.$$

Thus $f \circ p = r$, $p \circ f = q$.

$$(p \circ q)(x) = p(1-x) = \frac{1}{1-x}, \quad (q \circ p)(x) = q(1/x) = 1 - \frac{1}{x};$$

thus $p \circ q = f$, $q \circ p = g$. The complete table is

$\circ$	i	f	g	p	q	r
i	i	f	g	p	q	r
f	f	g	i	r	p	q
g	g	i	f	g	r	p
p	p	q	r	i	f	g
q	q	r	p	g	i	f
r	r	p	q	f	g	i

3.9 This follows immediately from the identity $\cos^2 \theta + \sin^2 \theta = 1$.

3.10 From the addition formula, $\cos 2\theta = \cos^2 \theta - \sin^2 \theta = \cos^2 \theta - (1 - \cos^2 \theta) = 2\cos^2 \theta - 1$. Hence $2\cos^2 \theta = 1 + \cos 2\theta$, and the result follows. Similarly, $\cos 2\theta = (1 - \sin^2 \theta) - \sin^2 \theta = 1 - 2\sin^2 \theta$, and so $\sin^2 \theta = (1/2)(1 - \cos 2\theta)$.

3.11 From the addition formulae, $\sin(\theta + \phi) + \sin(\theta - \phi) = 2\sin\theta\cos\phi$. Now let $x = \theta + \phi$, $y = \theta - \phi$. Then $\theta = (x+y)/2$, $\phi = (x-y)/2$, and the result follows. Similarly $\cos(\theta + \phi) + \cos(\theta - \phi) = 2\cos\theta\cos\phi$, and the result follows, with x and y defined as before.

3.12
$$\tan(x+y) = \frac{\sin(x+y)}{\cos(x+y)} = \frac{\sin x \cos x + \cos x \sin y}{\cos x \cos y - \sin x \sin y}.$$

Dividing the numerator and denominator by $\cos x \cos y$ gives the result.

3.13 $(1-t^2)/(1+t^2) = [\cos^2(\theta/2) - \sin^2(\theta/2)]/[\cos^2(\theta/2) + \sin^2(\theta/2)] = \cos\theta$, $2t/(1+t^2) = [2\cos(\theta/2)\sin(\theta/2)]/[\cos^2(\theta/2) + \sin^2(\theta/2)] = \sin\theta$, $\tan\theta = \sin\theta/\cos\theta = 2t/(1-t^2)$.

3.14 Suppose that f is increasing and bounded above. Then the set $\{f(x) : x \in [a, \infty)\}$, being bounded above, has a supremum A. For all $\epsilon > 0$ there exists X such that $f(X) > A - \epsilon$, since otherwise $A - \epsilon$ would be an upper bound for f. Since f is increasing, it follows that $A - \epsilon < f(x) \le A$ for all $x > X$. Thus $\lim_{x \to \infty} f(x) = A$.

3.15 $\lim_{x\to\infty} f(-x) = L$ if and only if for all $\epsilon > 0$ there exists $M > 0$ such that $|f(-x) - L| < \epsilon$ for all $x > M$, that is (writing $-x$ as y) if and only if for all $\epsilon > 0$ there exists $M > 0$ such that $|f(y) - L| < \epsilon$ for all $y < -M$.

3.16 Let f, defined on $[a, b]$, be increasing. Then $f(a) \le f(x) \le f(b)$ for every x in $[a, b]$, and so f is bounded both above and below. The result does not hold for an open interval: if $f(x) = 1/(1 - x)$ $(x \in [0, 1))$, then f is increasing, but is not bounded above.

3.17 Since $B \subseteq A$, we have that $\{f(x) : x \in B\} \subseteq \{f(x) : x \in A\}$. It follows that every upper bound of $\{f(x) : x \in A\}$ is also an upper bound of $\{f(x) : x \in B\}$. In particular $\sup_A f$ is an upper bound of $\{f(x) : x \in B\}$, and so $\sup_A f \ge \sup_B f$. A similar argument regarding lower bounds gives that $\inf_A f \le \inf_B f$.

3.18 Let f, with domain $[0, 1]$, be given by $f(x) = 1$ if x is rational, and $f(x) = -1$ if x is irrational. Then $\lim_{x\to 1} |f(x)| = 1$, but $\lim_{x\to 1} f(x)$ does not exist.

3.19 This follows from Theorem 3.3 and the observation that $\max\{f, g\} = (1/2)(f + g + |f - g|)$ and $\min\{f, g\} = (1/2)(f + g - |f - g|)$.

3.20 The identity $1 - \cos x = 2\sin^2(x/2)$ follows immediately from Exercise 3.10. Then

$$\frac{1 - \cos x}{x^2} = \frac{1}{2}\left(\frac{\sin(x/2)}{x/2}\right)^2 \to \frac{1}{2} \text{ as } x \to 0.$$

3.21 This follows from Theorem 3.11, since $\max\{f, g\} = (1/2)(f + g + |f - g|)$ and $\min\{f, g\} = (1/2)(f + g - |f - g|)$.

3.22 $(x - a)(b - x) \ge 0$ for all x in $[a, b]$. Since both the functions $x \mapsto (x-a)(b-x)$ and $x \mapsto \sqrt{x}$ are continuous, the function $x \mapsto \sqrt{(x - a)(b - x)}$ is continuous in $[a, b]$. The same applies to the function $x \mapsto \sqrt{(x - a)/(b - x)}$, except that the region of definition and of continuity is the interval $[a, b)$.

3.23 The function $x \mapsto \cot x = \cos x / \sin x$ is continuous except when $\sin x = 0$, that is, for x in $\mathbb{R} \setminus \{n\pi : n \in \mathbb{Z}\}$.

3.24 Let $\epsilon = |f(c)|/2$. Then there exists $\delta > 0$ such that $\left| |f(x)| - |f(c)| \right| \le |f(x) - f(c)| < |f(c)|/2$ for all x in $(c - \delta, c + \delta)$. It follows that $|f(c)|/2 < |f(x)| < 3|f(c)|/2$, and so certainly $f(x) \ne 0$ for all x in $(c - \delta, c + \delta)$.

3.25 Let $\epsilon > 0$ be given, and let $\delta = \epsilon$. There is no harm in supposing that $\epsilon < 1$. If $0 < x < \delta$, then

$$|f(x) - f(0)| = \begin{cases} x < \epsilon & \text{if } x \text{ is rational} \\ x^2 < \epsilon^2 < \epsilon & \text{if } x \text{ is irrational.} \end{cases}$$

Thus f is continuous at 0. Similarly, for continuity at 1, define $\delta = \epsilon/2$. Also, observe that, for all x in $(0,1)$, $1 - x^2 = (1-x)(1+x) < 2(1-x)$. Then, if $|1 - x| < \delta$, it follows that

$$|f(x) - f(1)| = \begin{cases} 1 - x < \delta < \epsilon & \text{if } x \text{ is rational} \\ 1 - x^2 < 2(1-x) < 2\delta = \epsilon & \text{if } x \text{ is irrational.} \end{cases}$$

Thus f is continuous at 1. Suppose finally that $0 < x < 1$, so that $x^2 \neq x$. If x is rational, and if, for all $n \geq 1$, we define $a_n = x + (\sqrt{2}/n)$, then $\lim_{n \to \infty} a_n = x$, whereas $\lim_{n \to \infty} f(a_n) = \lim_{n \to \infty} (a_n)^2 = x^2 \neq f(x)$. If x is irrational, we define b_n to be any chosen rational number in the interval $(x, x + (1/n))$. Then $\lim_{n \to \infty} b_n = x$, whereas $\lim_{n \to \infty} f(b_n) = \lim_{n \to \infty} b_n = x \neq f(x)$. From Theorem 3.5, we deduce that, whether x is rational or irrational, f is not continuous at x.

3.26 Let $g(x) = f(x) - x$. Then g is continuous on $[a,b]$. Since $a \leq f(x) \leq b$ for all x in $[a,b]$, we have $g(a) = f(a) - a \geq 0$ and $g(b) = f(b) - b \leq 0$. If either of these inequalities is an equality then the result is clear, with $c = a$ or $c = b$. Otherwise, by the intermediate value theorem (Theorem 3.12) there exists c in (a,b) such that $g(c) = 0$.

3.27 The function $f - g$ is continuous, and $(f-g)(0) < 0$, $(f-g)(1) > 0$. Hence, by the intermediate value theorem, there exists c in (a,b) such that $(f-g)(c) = 0$.

3.28 Suppose that f is not monotonic. Then there exist a, b, c in $\mathbb{R}$ such that $a < b < c$ and either (i) $f(a) < f(b)$, $f(b) > f(c)$, or (ii) $f(a) > f(b)$, $f(b) < f(c)$. Consider Case (i), and suppose first that $f(a) \leq f(c) < f(b)$. Let $d \in \big(f(c), (f(b)\big) \subseteq \big(f(a), f(b)\big)$. Then by the intermediate value theorem (Theorem 3.12) we obtain a contradiction, since there exist c_1 in (a,b) and c_2 in (b,c) such that $f(c_1) = f(c_2) = d$. Similarly, if we suppose that $f(c) \leq f(a) < f(b)$, then, for all d in $\big(f(a), (f(b)\big) \subseteq \big(f(c), f(b)\big)$, there exist c_1 in (a,b) and c_2 in (b,c) such that $f(c_1) = f(c_2) = d$. Case (ii) is dealt with in a similar way.

3.29 Consider the sequence

$$(x^{2^n}) = (x, x^2, x^4, x^8, \ldots),$$

where $x \in (-1,1)$. Then $(x^{2^n}) \to 0$, and so, by continuity at 0, the sequence $\big(f(x^{2^n})\big) \to f(0)$. But $\big(f(x^{2^n})\big)$ is the constant sequence $\big(f(x)\big)$, with limit $f(x)$, and so we conclude that $f(x) = f(0)$ for all x in $(-1,1)$.

3.30 Putting $x = 0$ in the functional equation, we find that $f(0) = f(a0) = bf(0)$, where $b > 1$, and this gives a contradiction unless $f(0) = 0$. Let

$|f(x)| \leq M$ for all x in $[0,1]$. Let $\epsilon > 0$ be given, and let $n \in \mathbb{N}$. Then $f(a^n x) = b^n f(x)$ for all x in $[0, a^{-n}]$, and so there exists N in $\mathbb{N}$ such that

$$|f(x) - f(0)| = |f(x)| = b^{-n}|f(a^n x)| \leq M b^{-n} < \epsilon$$

if $n > N$. Hence, defining δ as a^{-N}, we see that $|f(x) - f(0)| < \epsilon$ for all x in $(-\delta, \delta) \cap \operatorname{dom} f$. Thus f is continuous at 0.

3.31 For example, suppose that f and g are continuous at the point a. Then for every sequence (a_n) with limit a, we have that $(f(a_n)) \to f(a)$, $(g(a_n)) \to g(a)$. By Theorem 2.8, $((f+g)(a_n)) \to (f+g)(a)$, and so, by Theorem 3.6, $f + g$ is continuous at a.

3.32 Let $c \in \operatorname{dom} f$, and let (c_n) be a sequence contained in $\operatorname{dom} f$ such that $(c_n) \to c$. Then, using Theorem 3.5 twice, we deduce that $(f(c_n)) \to f(c)$ and $((g \circ f)(c_n)) \to (g \circ f)(c)$. Hence, by Theorem 3.6, $g \circ f$ is continuous.

3.33 For each $\epsilon > 0$ there exists $\delta_1 > 0$ such that $|f(x) - f(y)| < \epsilon/2$ for all x, y in $[a, b]$ such that $|x - y| < \delta_1$, and there exists $\delta_2 > 0$ such that $|f(x) - f(y)| < \epsilon/2$ for all x, y in $[b, c]$ such that $|x - y| < \delta_1$. Let $\delta = \min \{\delta_1, \delta_2\}$, and let x, y in $[a, c]$ be such that $|x - y| < \delta$. If both x and y are in $[a, b]$, or if both x and y are in $[b, c]$, then it is clear that $|f(x) - f(y)| < \epsilon/2 < \epsilon$. For the remaining case we may suppose without essential loss of generality that $x < b < y$. Then $|b - x| < \delta \leq \delta_1$ and so $|f(b) - f(x)| < \epsilon/2$. Similarly, $|y - b| < \delta \leq \delta_2$, and so $|f(y) - f(b)| < \epsilon/2$. Hence $|f(x) - f(y)| = |(f(x) - f(b)) + (f(b) - f(y))| \leq |f(b) - f(x)| + |f(y) - f(b)| < \epsilon$.

3.34 Let $\sin^{-1} x = \alpha$, so that $\alpha \in [0, \pi/2]$ and $\sin \alpha = x$. Then $\cos((\pi/2) - \alpha) = x$, and $(\pi/2) - \alpha \in [0, \pi]$. It follows that $\cos^{-1} x = (\pi/2) - \alpha$.

3.35 As an extreme case, consider the constant function C_k, defined by the rule that $C_k(x) = k$ for all x. The image of the function is $\{k\}$, but there is no inverse function from $\{k\}$ to $\mathbb{R}$.

3.36 Since $8 + 2x - x^2 = 9 - (x-1)^2$, the function has maximum value 9, obtained when $x = 1$. Now $y = 8 + 2x - x^2 \iff x^2 - 2x + (y - 8) = 0 \iff x = 1 \pm \sqrt{9 - y}$. There is an inverse function $f^{-1} : (-\infty, 9] \to [1, \infty)$ given by $f^{-1}(y) = 1 + \sqrt{9 - y}$.

3.37 The image of f is $\mathbb{R} \backslash \{0\}$, and this is the domain of the inverse function. The formula is obtained by observing that $y = 1/(1 - x) \iff x = 1 - (1/y)$. So $f^{-1}(y) = 1 - (1/y) \quad (y \in \mathbb{R} \backslash \{1\})$.

Chapter 4

4.1 It is easy to see that $f(x) \to 0 = f(0)$ as $x \to 0$, and so f is continuous at 0. However,

$$\frac{f(x) - f(0)}{x - 0} = \frac{\sqrt{|x|}}{x} = \begin{cases} 1/\sqrt{x} & \text{if } x > 0 \\ -1/\sqrt{|x|} & \text{if } x < 0, \end{cases}$$

and so f is not differentiable at 0.

4.2 a) Suppose that $\lim_{x \to c}[(f(x) - f(c))/(x - c)] = f'(c)$. Then for every $\epsilon > 0$ there exists $\delta > 0$ such that

$$\left| \frac{f(x) - f(c)}{x - c} - f'(c) \right| < \epsilon \tag{9.16}$$

for all x in dom $f \setminus \{c\}$ such that $|x - c| < \delta$. Then (9.16) holds for all x in dom $f \cap (-\infty, c)$ such that $|x - c| < \epsilon$, and it also holds for x in dom $f \cap (c, \infty)$ such that $|x - c| < \epsilon$. Thus the left derivative and right derivative both exist, and both are equal to $f'(c)$.

b) If $x < 0$ then $x(x-1) > 0$, and so $(f(x) - f(0))/(x-0) = x(x-1)/x = x - 1 \to -1$ as $x \to 0-$. If $x > 0$, and is sufficiently close to 0, then $x(x-1) < 0$, and so $(f(x) - f(0))/(x-0) = -x(x-1)/x = -x+1 \to 1$ as $x \to 0+$. Thus $f_l'(x) = -1$, $f_r'(x) = 1$.

4.3 Since $\lim_{x \to 0}(f(x) - f(0))/(x - 0) = x\sin(1/x) \to 0$ as $x \to 0$, we have that $f'(0) = 0$. If $x \neq 0$ then, by ordinary calculus methods, $f'(x) = 2x\sin(1/x) - \cos(1/x)$, and this does not have a limit as $x \to 0$. Thus f' is not continuous at 0.

4.4 No: if $f(x) = x$ and $g(x) = -x$, then $\max\{f, g\}$ and $\min\{f, g\}$ are not differentiable at 0.

4.5 Let $n = -m$, where m is a positive integer. Then $f(x) = 1/x^m$, and so $f'(x) = (-1/x^{2m})mx^{m-1} = (-m)x^{-m-1} = nx^{n-1}$.

4.6 Note that $D_x(\tan x) = \tan^2 x + 1$. By L'Hôpital's rule,

$$\lim_{x \to 0} \frac{\tan x - x}{x^3} = \lim_{x \to 0} \frac{\tan^2 x}{3x^2} = \lim_{x \to 0} \frac{2\tan x + 2\tan^3 x}{6x}$$

$$= \lim_{x \to 0} \frac{2 + 8\tan^2 x + 6\tan^4 x}{6} = \frac{1}{3}.$$

Again, by L'Hôpital's rule,

$$\lim_{x \to 0} \frac{\sin x - x\cos x}{x^3} = \lim_{x \to 0} \frac{x\sin x}{3x^2} = \lim_{x \to 0} \frac{\sin x}{3x} = \frac{1}{3}.$$

4.7 a) If $\alpha > 0$, then $|f(x) - f(a)| \to 0$ as $x \to a$, and so f is continuous at a. If $\alpha > 1$, then $|(f(x) - f(a))/(x - a)| < M|x - a|^{\alpha - 1}$, and so $f'(a) = 0$.

   b) Let $f(x) = |x|$. Then $|f(x) - f(0)| = \big||x|\big| = |x| = |x - 0|$, and so f, which is not differentiable at 0, nonetheless satisfies a Hölder condition in which $M = 2$ and $\alpha = 1$.

4.8 Let $x \in [a, b]$. If $x > c$, then $f(x) \leq f(c)$, and so $[f(x) - f(c)]/(x - c) \leq 0$. Hence $\lim_{x \to c+}[f(x) - f(c)]/(x - c) \leq 0$. Similarly $\lim_{x \to c-} f(x) - f(c)]/(x - c) \geq 0$. Since f is differentiable, the two limits are both equal to $f'(c)$, and so $f'(c) = 0$. The proof for d is similar.

4.9 For every c in the domain, $|(f(x) - f(c))/(x - c)| < |x - c|$, and so $f'(c) = 0$. It follows that f is constant.

4.10 We show that $\big(f(1/n)\big)$ is a Cauchy sequence. Let $\epsilon > 0$ be given, and choose an integer N such that $1/N < \epsilon$. Let $m, n > N$. Then, for some c between $1/m$ and $1/n$, $|f(1/m) - f(1/n)| = |(1/m) - (1/n)| \, |f'(c)| < |(1/m) - (1/n)| < 1/N < \epsilon$.

4.11 By Theorem 4.15 $(\cos^{-1})'(y) = 1/[-\sin(\cos^{-1} y)]$. Since $0 < \cos^{-1} y < \pi$, it follows that $\sin(\cos^{-1} y)$, being positive, is equal to $\sqrt{1 - \cos^2(\cos^{-1} y)} = \sqrt{1 - y^2}$. Hence $(\cos^{-1})'(y) = -1/\sqrt{1 - y^2}$ for all y in $(-1, 1)$.

4.12 Since $D_x(\cos^{-1} x + \sin^{-1} x) = 0$, it follows that $\cos^{-1} x + \sin^{-1} x$ is constant. Its value at 0 is $\cos^{-1} 0 + \sin^{-1} 0 = \pi/2 + 0 = \pi/2$, and so this is its value throughout $[-1, 1]$.

4.13 Since $f'(x) = f(x) > 0$ for all x, it follows that f is an increasing function. By Theorem 4.15, $(f^{-1})'(x) = 1/f'(f^{-1}(x)) = 1/f(f^{-1}(x)) = 1/x$.

4.14 We use L'Hôpital's rule twice, differentiating with respect to h: thus, $\lim_{h \to 0}(1/h^2)[f(a + 2h) - 2f(a + h) + f(a)] = \lim_{h \to 0}(1/2h)[2f'(a + 2h) - 2f'(a + h)] = \lim_{h \to 0}(1/2)[4f''(a + 2h) - 2f''(a + h)] = f''(a)$.

4.15 Denote $x^2 \cos x$ by $h(x)$. Observe that $\cos^{(2k-1)} x = (-1)^k \sin x$, $\cos^{(2k)} x = (-1)^k \cos x$. Then, by Leibniz's Theorem, $h^{2n-1}(x) = (-1)^{n-1}\big[((2n - 1)(2n - 2) - x^2)\sin x + 2(2n - 1)x \cos x\big]$.

4.16 Since $f'(x) = -m\sin(m \sin^{-1} x)/\sqrt{1 - x^2}$, if follows that $\sqrt{1 - x^2}f'(x) = -m\sin(m \sin^{-1} x)$. Differentiating again, we obtain (after some easy algebra)
$$(1 - x^2)f''(x) - xf'(x) + m^2 f(x) = 0.$$
Hence the required result is true for $n = 0$. Now suppose inductively that
$$(1 - x^2)f^{(n+1)}(x) - (2n - 1)xf^{(n)}(x) + \big(m^2 - (n - 1)^2\big)f^{(n-1)}(x) = 0.$$

Then, differentiating, we have

$$[(1 - x^2)f^{(n+2)}(x) - 2xf^{(n+1)}(x)] - (2n - 1)[xf^{(n+1)}(x) + f^{(n)}(x)]$$
$$+ (m^2 - n^2 + 2n - 1)f^{(n)}(x) = 0,$$

and collecting terms gives us

$$(1 - x^2)f^{(n+2)}(x) - (2n + 1)xf^{(n+1)}(x) + (m^2 - n^2)f^{(n)}(x) = 0.$$

Hence, by induction, the result holds for all $n \geq 0$.

Now put $x = 0$ to obtain $f^{(n+2)}(0) = (n^2 - m^2)f^{(n)}(0)$ for all $n \geq 0$. Since $f(0) = 1$ and $f'(0) = 0$, we see that $f^{(n)}(0) = 0$ for all odd n, while for even n we have $f^{(n)}(0) = [(n - 2)^2 - m^2] \ldots [2^2 - m^2][-m^2]$.

4.17 If $f(x) = a_0 + a_1 x + \cdots + a_n x^n$, then $f^{(r)}(x) = r!a_r$ + positive powers of x for all $r < n$, $f^{(n)}(x) = n!a_n$, and $f^{(r)}(x) = 0$ for all $r > n$. Thus

$$\frac{1}{r!}f^{(r)}(0) = \begin{cases} a_r & \text{if } 0 \leq r \leq n \\ 0 & \text{if } r > n. \end{cases}$$

Chapter 5

5.1 In every subinterval of any dissection D, the supremum and infimum of the function is k, and so $\mathcal{U}(C_k, D) = \mathcal{L}(C_k, D) = k(b - a)$. If follows that $\int_a^b C_k = k(b - a)$.

5.2 Let $D_n = \{0, \frac{1}{n}, \frac{2}{n}, \cdots, \frac{n-1}{n}, 1\}$. Then, using Exercise 1.7, we see that $\mathcal{U}(f, D_n) = (1/n)\sum_{i=1}^n (i^2/n^2) = (1/n^3)(1^2 + 2^2 + \cdots + n^2) = (1/6)[1 + (1/n)][2 + (1/n)]$ while $\mathcal{L}(f, D_n) = (1/n^3)(1^2 + 2^2 + \cdots + (n-1)^2) = (1/6)[1 - (1/n)][2 - (1/n)]$. Hence, for all n, $\overline{\int_0^1} f - \underline{\int_0^1} f \leq \mathcal{U}(f, D_n) - \mathcal{L}(f, D_n) = 1/n$, and so f is Riemann integrable. Since $\int_0^1 f$ lies between the upper and lower sums, both of which have limit $1/3$ as $n \to \infty$. we also deduce that $\int_0^1 f = 1/3$.

5.3 Choose a dissection $D = \{x_0, x_1, \ldots, x_n\}$ containing c. Then, recalling that M_i and m_i are (respectively) the supremum and the infimum of f in the *open* subinterval (x_{i-1}, x_i), we see that $M_i = m_i = 0$ for all i, and so $\mathcal{U}(f, D) = \mathcal{L}(f, D) = 0$. Hence $\overline{\int_0^1} f - \underline{\int_0^1} f \leq \mathcal{U}(f, D) - \mathcal{L}(f, D) = 0$, and so $f \in \mathcal{R}[0, 1]$. Since $\int_0^1 f$ must lie between $\mathcal{U}(f, D)$ and $\mathcal{U}(f, D)$, we must have $\int_0^1 f = 0$.

5.4 Suppose for convenience that $c_1 < c_2 < \cdots < c_k$, and look first at the case where $c_1 > a$ and $c_k < b$ Let $\epsilon > 0$ be given. Define

$$\delta = \min\left\{c_1 - a, b - c_k, \frac{\epsilon}{4k(M - m)}\right\},$$

where $M = \sup_{[a,b]} f$, $m = \inf_{[a,b]} f$. Then let D be a partition in which the subinterval $(c_i - \delta, c_i + \delta)$ features, for $i = 1, 2, \ldots, k$. In each of the intervals $[a, c_1 - \delta]$, $[c_1 + \delta, c_2 - \delta]$, $\ldots$, $[c_{k-1} + \delta, c_k - \delta]$, $[c_k + \delta, b]$ the function is continuous, and so we can arrange for the contribution of those intervals to $\mathcal{U}(f, D) - \mathcal{L}(f, D)$ to be less than $\epsilon/2$. The contribution of each of the intervals $[c_i - \delta, c_i + \delta]$ cannot exceed $2\delta(M - m) = \epsilon/2k$, and so the total contribution of those intervals is at most $\epsilon/2$. Hence $\mathcal{U}(f, D) - \mathcal{L}(f, D) < (\epsilon/2) + (\epsilon/2) = \epsilon$, and so $f \in \mathcal{R}[a, b]$. Small and straightforward adjustments in the argument are necessary to cope with the case where $c_1 = a$ and/or $c_k = b$.

5.5 Let $\epsilon = f(c)/2$. By continuity, there exists $\delta > 0$ such that $|f(x) - f(c)| < \epsilon$ for all x in $[a, b]$ such that $|x - c| < \delta$. It follows that $f(c) - (f(c)/2) < f(x) < f(c) + (f(c)/2)$, and so in particular $f(x) > f(c)/2$ for all x in the interval $[a, b] \cap (c - \delta, c + \delta)$, an interval whose length is at least δ. Thus $\int_a^b f \geq \delta(f(c)/2) > 0$.

5.6 The function $x \mapsto (f(x))^2$ is continuous and takes only non-negative values. By the previous exercise, $\int_a^b (f(x))^2\, dx = 0$ implies that $(f(x))^2 = 0$ for all x in $[a, b]$.

5.7 Let D be a dissection of $[a, b]$ containing c and d, and let $D_1 = D \cap [a, c]$, $D_2 = D \cap [c, d]$, $D_3 = D \cap [d, b]$. (The situation simplifies if $c = a$ or $d = b$.) Then $\mathcal{U}(f, D) - \mathcal{L}(f, D) = \sum_{i=1}^3 (\mathcal{U}(f, D_i) - \mathcal{L}(f, D_i))$. If, for a given $\epsilon > 0$, we choose D so that $\mathcal{U}(f, D) - \mathcal{L}(f, D) < \epsilon$, then certainly $\mathcal{U}(f, D_2) - \mathcal{L}(f, D_2) < \epsilon$, and so $f \in \mathcal{R}[c, d]$.

5.8 Let $\epsilon > 0$ be given. By uniform continuity, there exists $\delta > 0$ such that $|f(x) - f(y)| < \epsilon/2(b - a)$ for all x, y in $[a, b]$ such that $|x - y| < \delta$. Choose n so that $n > 1/\delta$; thus each subinterval in D_n has length less than δ. Then in each subinterval (x_{i-1}, x_i) there exists x such that $M_i - f(x) < \epsilon/4(b - a)$ and there exists y such that $f(y) - m_i < \epsilon/4(b - a)$. It follows that $M_i - m_i = (M_i - f(x)) + (f(x) - f(y)) + (f(y) - m_i) < \epsilon/4(b - a) + |f(x) - f(y)| + \epsilon/4(b - a) < \epsilon$, since $|x - y| < \delta$. It follows that $\mathcal{U}(f, D_n) - \mathcal{L}(f, D_n) = ((b - a)/n) \sum_{i=1}^n (M_i - m_i) < \epsilon$ if n is sufficiently large. Now $\mathcal{L}(f, D_n) \leq \int_a^b f \leq \mathcal{U}(f, D_n)$, and so certainly $|\mathcal{U}(f, D_n) - \int_a^b f| < \epsilon$ and $|\mathcal{L}(f, D_n) - \int_a^b f| < \epsilon$ for sufficiently large n. It follows that the sequences $(\mathcal{U}(f, D_n))$ and $(\mathcal{L}(f, D_n))$ both have limit $\int_a^b f$.

5.9 Let D_m be the dissection $\{0, 1/2^{2m}, 1/2^{2m-1}, \ldots, 1/2, 1\}$. In all but the first subinterval, the function is constant, and so the supremum and infimum are equal. In the first subinterval the length is $1/2^{2m}$, the supremum is $1/2^{2m}$ and the infimum is $-1/2^{2m+1}$. Hence $\mathcal{U}(f, D_m) - \mathcal{L}(f, D_m) = (1/2^{2m})[(1/2^{2m}) + (1/2^{2m+1})] = 3/2^{4m+1}$, and this can be made less than any given ϵ by taking m sufficiently large. Hence $f \in \mathcal{R}[0, 1]$. The value of the integral is the limit of $\mathcal{U}(f, D_m)$, namely the sum of the geometric series $1 \cdot \frac{1}{2} - \frac{1}{2} \cdot \frac{1}{4} + \frac{1}{4} \cdot \frac{1}{8} - \cdots$. The sum is $\frac{1}{2} / \left(1 + \frac{1}{4}\right) = \frac{2}{5}$.

5.10 Let $f(x) = g(x) = x$. Then $\int_0^1 f \cdot g = \int_0^1 x^2\, dx = \frac{1}{3}$, but $\left(\int_0^1 f\right)\left(\int_0^1 g\right) = \frac{1}{4}$.

5.11 Properties a) and b) follow immediately from Theorem 5.15. It follows from Theorem 5.14 that $(f, f) \geq 0$, and from Exercise 5.6 that $(f, f) = 0$ only if $f = 0$. The inner product $(kf + g, kf + g)$ is non-negative for all k. That is, $k^2(f, f) + 2k(f, g) + (g, g) = 0$ for all k. It follows that the discriminant is non-positive: $4(f, g)^2 - 4(f, f)(g, g) \leq 0$, and so $\left(\int_a^b f \cdot g\right)^2 \leq \left(\int_a^b f^2\right)\left(\int_a^b g^2\right)$.

5.12 Let $f(x) = 0$ for $a \leq x \leq \frac{1}{2}(a + b)$, and $f(x) = 1$ for $\frac{1}{2}(a + b) < x \leq b$. Then $\int_a^b f = \frac{1}{2}(b - a)$, but there is no c such that $f(c) = \frac{1}{2}$.

5.13 Since $g(x) \geq 0$, we have that $mg(x) \leq f(x)g(x) \leq Mg(x)$ for all x in $[a, b]$, where $M = \sup_{[a,b]} f$, $m = \inf_{[a,b]} f$. Thus $m \int_a^b g \leq \int_a^b f \cdot g \leq M \int_a^b g$, and so, since $\int_a^b g > 0$, $m \leq \left(\int_a^b f \cdot g\right) / \left(\int_a^b g\right) \leq M$. By the intermediate value theorem, there exists c in $[a, b]$ such that $f(c) = \left(\int_a^b f \cdot g\right) / \left(\int_a^b g\right)$.

5.14 Since $f(x) = \frac{1}{2}x^2 \int_0^x g(t)\, dt - x \int_0^x tg(t)\, dt + \frac{1}{2} \int_0^x t^2 g(t)\, dt$, we may deduce that $f'(x) = \frac{1}{2}x^2 g(x) + x \int_0^x g(t)\, dt - x^2 g(x) - \int_0^x tg(t)\, dt + \frac{1}{2}x^2 g(x) = x \int_0^x g(t)\, dt - \int_0^x tg(t)\, dt$. Hence $f''(x) = \int_0^x g(t)\, dt + xg(x) - xg(x) = \int_0^x g(t)\, dt$, and $f'''(x) = g(x)$.

5.15 $f(x) = g(u) - g(v)$, where $u = x^2$, $v = x^3$, and $g(u) = \int_0^u t^5/(1 + t^4)\, dt$. Hence $f'(x) = [u^5/(1 + u^4)] \cdot 2x - [v^5/(1 + v^4)] \cdot 3x^2 = 2x^{11}/(1 + x^8) - 3x^{17}/(1 + x^{12})$.

5.16
$$\int_0^{\pi/4} \tan^2 x\, dx = \left[\tan x - x\right]_0^{\pi/4} = 1 - (\pi/4);$$

$$\int_0^{\pi/2} \sin^2 x\, dx = \frac{1}{2} \int_0^{\pi/2} (1 - \cos 2x)dx = \frac{1}{2}\left[x - \frac{1}{2}\sin 2x\right]_0^{\pi/2} = \frac{1}{4}\pi^2.$$

5.17 The argument takes no account of the indefiniteness of an indefinite integral.

5.18 $I_0 = \int_0^{\pi/2} x\, dx = \pi^2/8$, $I_1 = \Big[x(-\cos x)\Big]_0^{\pi/2} - \int_0^{\pi/2}(-\cos x)dx = 0 +$
$\Big[\sin x\Big]_0^{\pi/2} = 1$. To obtain the reduction formula, integrate by parts, taking the factors as $\sin x$ and $x\sin^{n-1}x$:

$$I_n = \int_0^{\pi/2} x\sin^n x\, dx = \Big[(-\cos x)x\sin^{n-1}x\Big]_0^{\pi/2}$$

$$+ \int_0^{\pi/2}(\cos x)[\sin^{n-1}x + (n-1)x\sin^{n-2}x\cos x]\,dx$$

$$= 0 + \int_0^{\pi/2}\sin^{n-1}x\cos x\, dx + (n-1)\int_0^{\pi/2}x\sin^{n-2}x(1-\sin^2 x)\,dx$$

$$= \Big[\frac{1}{n}\sin^n x\Big]_0^{\pi/2} + (n-1)I_{n-2} - (n-1)I_n$$

$$= \frac{1}{n} + (n-1)I_{n-2} - (n-1)I_n .$$

Hence
$$I_n = \frac{1}{n^2} + \frac{n-1}{n}I_{n-2} .$$
So $I_3 = \frac{1}{9} + \frac{2}{3}I_1 = \frac{7}{9}$, and $I_4 = \frac{1}{16} + \frac{3}{4}(\frac{1}{4} + \frac{1}{2}I_0) = \frac{1}{4} + \frac{3}{64}\pi^2$.

5.19

$$\int_0^1 \frac{x^5\, dx}{\sqrt{1+x^6}} = \frac{1}{6}\int_1^2 \frac{du}{\sqrt{u}} \text{ (with } u = 1+x^6) = \frac{1}{3}\Big[\sqrt{u}\Big]_1^2 = \frac{1}{3}(\sqrt{2}-1).$$

$$\int_0^{\pi/2} \frac{\sin x\, dx}{(3+\cos x)^2} = \int_4^3 \frac{-du}{u^2} \text{ (with } u = 3+\cos x) = \frac{1}{12}.$$

$$\int_0^{\pi/4} \cos 2x\sqrt{4-\sin 2x} = -\frac{1}{2}\int_4^3 \sqrt{u}\, du \text{ (with } u = 4-\sin 2x) = \frac{1}{3}(8-3\sqrt{3}).$$

5.20 Since $\sin(\pi - x) = \sin x$, putting $u = \pi - x$ gives $I = \int_0^\pi xf(\sin x)\,dx =$
$\int_\pi^0 (\pi - u)f\sin(\pi - u)(-du) = \pi\int_0^\pi f(\sin u)\,du - \int_0^\pi uf(\sin u)\,du$
$= \pi\int_0^\pi f(\sin x)\,dx - I$, and the result follows. Since $\cos^2 x = 1 - \sin^2 x$,
we can apply the result to the given integral I, making the substitution
$u = \cos x$, obtaining

$$I = \frac{\pi}{2}\int_0^\pi \frac{\sin x\, dx}{1+\cos^2 x} = \frac{\pi}{2}\int_1^{-1}\frac{-du}{1+u^2} = \frac{\pi}{2}\Big[\tan^{-1}u\Big]_{-1}^1 = \frac{\pi^2}{4}.$$

5.21 The result holds for $n = 1$: $f(x) = f(a) + \int_a^x f'(t)\,dt$. For the inductive step, note that integration by parts gives

$$R_n = \frac{1}{(n-1)!} \int_a^x (x-t)^{n-1} f^{(n)}(t)\,dt$$

$$= \frac{1}{(n-1)!}\left[-\frac{(x-t)^n}{n}f^{(n)}(t)\right]_{t=a}^{t=x} - \frac{1}{(n-1)!}\int_a^x -\frac{(x-t)^n}{n}f^{(n+1)}(t)\,dt$$

$$= \frac{(x-a)^n}{n!}f^{(n)}(a) + R_{n+1}$$

and the result follows.

5.22 Since $x/\sqrt{x^6+1} \sim 1/x^2$ as $x \to \infty$, the first integral converges. Since $(2x+1)/(3x^2 + 4\sqrt{x} + 7) \asymp 1/x$ as $x \to \infty$, the second integral diverges.

5.23 Since

$$\frac{Kx}{x^2+1} - \frac{1}{2x+1} = \frac{(2K-1)x^2 + Kx - 1}{(x^2+1)(2x+1)} \asymp \begin{cases} 1/x & \text{if } K \neq 1/2 \\ 1/x^2 & \text{if } K = 1/2, \end{cases}$$

the integral converges if and only if $K = 1/2$.

Denote the second integrand by $F(x)$. Then

$$F(x) = \frac{(x+1)^2 - K^2(2x^2+1)}{(x+1)\sqrt{2x^2+1}(x+1+K\sqrt{2x^2+1}}$$

$$= \frac{x^2(1 - 2K^2) + 2x + (1-K)}{(x+1)\sqrt{2x^2+1}(x+1+K\sqrt{2x^2+1}}$$

$$\asymp \begin{cases} 1/x & \text{if } K \neq \pm 1/\sqrt{2} \\ 1/x^2 & \text{if } K = \pm 1/\sqrt{2}. \end{cases}$$

Thus the integral converges if and only if $K = \pm 1/\sqrt{2}$.

5.24 For $0 \leq x \leq 1$, let $g(x) = x(1-x)$, and let $f(x) = g(x - \lfloor x \rfloor)$ for $x \geq 0$. Thus the graph of g repeats between any two positive integers, and $f(n) = 0$ for $n = 0, 1, 2, \ldots$. Trivially, $\sum_{n=1}^\infty f(n)$ is convergent, but, for $N \in \mathbb{N}$, $\int_1^N f = (N-1)\int_0^1 x(1-x)\,dx = (N-1)/6$, and so $\int_0^\infty f$ diverges.

5.25 a) Integration by parts gives $\int_1^K (\sin x/x)\,dx = \left[-\cos x/x\right]_1^K - \int_1^K (\cos x/x^2)\,dx$. Since $\int_1^\infty (\cos x/x^2)\,dx$ is (absolutely) convergent, and since $\cos K/K \to 0$ as $K \to \infty$, the integral is convergent.

b) In the interval $[2k\pi, (2k+1)\pi]$, $\sin x \geq 0$. So $\int_{2k\pi}^{(2k+1)\pi} |\sin x/x|\,dx = \int_{2k\pi}^{(2k+1)\pi} (\sin x/x)\,dx \geq (1/(2k+1)\pi)\int_{2k\pi}^{(2k+1)\pi} \sin x\,dx = 2/(2k+1)\pi$. In the interval $[(2k-1)\pi, 2k\pi]$, $\sin x \leq 0$. So $\int_{(2k-1)\pi}^{2k\pi} |\sin x/x|\,dx = \int_{(2k-1)\pi}^{2k\pi} (-\sin x/x)\,dx \geq (1/2k\pi)\int_{(2k-1)\pi}^{2k\pi} (-\sin x))\,dx = 2/2k\pi$.

Hence $\int_{\pi}^{2N\pi} |\sin x/x|\,dx = \int_{\pi}^{2\pi} |\sin x/x|\,dx + \int_{2\pi}^{3\pi} |\sin x/x|\,dx + \cdots + \int_{(2N-1)\pi}^{2N\pi} |\sin x/x|\,dx \geq (2/\pi)\sum_{r=2}^{2N}(1/r)$, and by the divergence of the harmonic series it follows that I is not absolutely convergent.

5.26 $1/\sqrt{\sin x} \sim 1/\sqrt{x}$ as $x \to 0+$. Hence the first integral converges. Since $\sin x/x^2 \sim 1/x$ as $x \to 0+$, the second integral diverges. Since, by L'Hôpital's rule, $(x - \sin x)/x^3 \to 1/6$ as $x \to 0$, it follows that $1/(x - \sin x) \asymp 1/x^3$ as $x \to 0+$. Hence the integral is divergent.

5.27 Since $\sin x/x \to 1$ as $x \to 0+$, there is no problem regarding the lower limit. We have seen that the first integral converges, but is not absolutely convergent. The second integral is easily seen to be absolutely convergent.

$$\int_0^\infty \frac{\sin^2 x}{x^2}\,dx = \left[\sin^2 x \left(-\frac{1}{x}\right)\right]_0^\infty - \int_0^\infty \left(-\frac{1}{x}\right) 2\sin x \cos x\,dx$$

$$= 0 + \int_0^\infty \frac{\sin 2x}{x}\,dx = \int_0^\infty \frac{\sin u}{u}\,du\,, \text{ where } u = 2x.$$

Chapter 6

6.1 $L(2^n) = n\log 2$, and $\log 2 = \int_1^2 (dx/x) > 1/2$, since $1/2 = \inf\{1/x : 1 \leq x \leq 2\}$.

6.2 $(1-u)(1+u) = 1-u^2 < 1$ and so, if u is positive, $1-u < 1/(1+u) < 1$. It follows that, for all positive x, $\int_0^x (1-u)\,du < \int_0^x (1/(1+u))\,du < \int_0^x du$; that is, $x - \frac{1}{2}x^2 < \log(1+x) < x$.

6.3 Observe first that $f(1) = 0$. Now, $f'(x) = 1 - (1/x)$, and so $f'(x) > 0$ if $0 < x < 1$, and $f'(x) < 0$ if $x > 1$. By the mean value theorem, $f(x) = f(1) + (x-1)f'(c) = (x-1)f'(c)$, where c is between 1 and x. Since $(x-1)f'(c)$ is positive for all $x \neq 1$, we have $\log x < x - 1$. Similarly, $g(1) = 0$, and $g'(x) = (1/x) - (1/x^2)$, which is negative if $x < 1$ and positive if $x > 1$. Thus $g(x) = (x-1)g'(c)$ (where c is between 1 and x) is positive for all $x \neq 1$, and so $1 - (1/x) < \log x$.

6.4 From Taylor's Theorem, $\log(1+x) = \log 1 + x(\log)'(1) + (x^2/2)(\log)''(1) + (x^3/6)(\log)'''(t)$, where $1 < t < x$. That is, $\log(1+x) = x + (x^2/2) + (x^3/6)(2/t^3) > x + (x^2/2)$. Again, we have $\log(1+x) = \log 1 + x(\log)'(1) + (x^2/2)(\log)''(1) + (x^3/6)(\log)'''(1) + (x^4/24)(\log)^{(4)}(t) = x + (x^2/2) + (x^3/3) - (x^4/24)(6/t^4) < x + (x^2/2) + (x^3/3)$.

6.5 Using integration by parts, we have

$$L(m,n) = \frac{x^{m+1}(\log x)^n}{m+1} - \int \frac{x^{m+1}}{m+1} n(\log x)^{n-1} \frac{1}{x} \, dx$$

$$= \frac{x^{m+1}(\log x)^n}{m+1} - \frac{n}{m+1} L(m, n-1),$$

and so (noting that $L(1,0) = x^2/2$) we have

$$L(1,3) = \frac{x^2}{2}(\log x)^3 - \frac{3}{2} L(1,2) = \frac{x^2}{2}(\log x)^3 - \frac{3}{2}\left[\frac{x^2}{2}(\log x)^2 - L(1,1)\right]$$

$$= \frac{1}{2}x^2(\log x)^3 - \frac{3}{4}x^2(\log x)^2 + \frac{3}{2}\left[\frac{x^2}{2}\log x - \frac{x^2}{2}\right]$$

$$= \frac{1}{4}x^2\left(2(\log x)^3 - 3(\log x)^2 + 3\log x - 3\right).$$

6.6 By L'Hôpital's rule,

$$\lim_{x\to 0} \frac{\log(\cos ax)}{\log(\cos bx)} = \lim_{x\to 0} \frac{-a\sin ax/\cos ax}{-b\sin bx/\cos bx} = \frac{a}{b} \cdot \lim_{x\to 0} \frac{\sin ax}{\sin bx} \cdot \lim_{x\to 0} \frac{\cos bx}{\cos ax}$$

$$= \frac{a}{b} \cdot \lim_{x\to 0} \frac{a\cos ax}{b\cos bx} \cdot 1 = \frac{a^2}{b^2}.$$

6.7 By Taylor's Theorem, $e^{-x} = 1 - x + (x^2/2)e^{-\theta x}$, where $0 < \theta < 1$. Since both x^2 and $e^{-\theta x}$ are positive for all $x \neq 0$, it follows that $e^{-x} > 1 - x$. Replacing x by $-x$ gives $e^x > 1 + x$, and this holds for all $x \neq 0$. Taking reciprocals gives $e^x < 1/(1-x)$, provided $1 - x$ is positive.

6.8 a) $y \in \operatorname{im} \cosh$ if and only if there exists x such that $e^{2x} - 2ye^x + 1 = 0$. This is a quadratic equation in e^x, and so $e^x = y \pm \sqrt{y^2 - 1}$. Since e^x must be positive, a suitable x exists only if $y \geq 1$. Similarly, $y \in \operatorname{im} \sinh$ if and only if there exists x such that $e^{2x} - 2ye^x - 1 = 0$, that is, if and only if $e^x = y \pm \sqrt{1 + y^2}$. The appropriate x is $\log(y + \sqrt{1 + y^2})$, and exists for all y.

b) These are all a matter of routine algebra. For example, $\sinh x \cosh y + \cosh x \sinh y = (1/4)[(e^x - e^{-x})(e^y + e^{-y}) + (e^x + e^{-x})(e^y - e^{-y})] = (1/4)[e^{x+y} - e^{-x+y} + e^{x-y} - e^{-(x+y)} + e^{x+y} + e^{-x+y} - e^{x-y} - e^{-(x+y)}] = (1/2)(e^{x+y} - e^{-(x+y)}) = \sinh(x + y)$.

c) $\sinh$ has a positive derivative throughout its domain, and so has a inverse function $\sinh^{-1} : \mathbb{R} \to \mathbb{R}$ with positive derivative. Moreover

$$(\sinh^{-1})'(x) = \frac{1}{\cosh(\sinh^{-1} x)} = \frac{1}{\sqrt{1 + \left(\sinh(\sinh^{-1} x)\right)^2}} = \frac{1}{\sqrt{1 + x^2}}.$$

In $[0, \infty)$ the function cosh has a positive derivative, and so there is an inverse function $\cosh^{-1} : [0\infty) \to [1, \infty)$, with positive derivative. Also, $(\cosh^{-1})'(x) = 1/\sinh(\cosh^{-1}x) = 1/\sqrt{x^2 - 1}$.

d) This amounts to solving the equations $y = \sinh x$ and $y = \cosh x$ for x, something already done in part a).

6.9 Substitute $u = \log x$; then $du = dx/x$, and so $\int(dx/x\log x) = \int(du/u) = \log u = \log\log x$. Now $\log\log x \to \infty$ as $x \to \infty$, and consequently the integral $\int_2^\infty (dx/x\log x)$ is divergent. By the integral test (Theorem 5.37) so is the series $\sum_{n=3}^\infty (1/n\log n)$. Since, for all integers $k \geq 3$,

$$\frac{1}{(k+1)\log(k+1)} \leq \int_k^{k+1} \frac{dx}{x\log x} \leq \frac{1}{k\log k}$$

it follows that

$$\sum_{r=4}^n \frac{1}{r\log r} \leq \log\log n - \log\log 3 \leq \sum_{r=3}^{n-1} \frac{1}{r\log r},$$

from which it follows that

$$\frac{1}{n\log n} \leq \delta_n + \log\log 3 \leq \frac{1}{3\log 3}.$$

Since

$$\delta_n - \delta_{n-1} = \frac{1}{n\log n} - \int_{n-1}^n \frac{dx}{x\log x} \leq 0,$$

the decreasing sequence (δ_n) has a limit δ. Since $K_n \sim \log\log n$, we require $e^{e^5} \approx 2.85 \times 10^{64}$ terms for the sum to exceed 5.

6.10 The same method can be used to compare the series

$$\sum_{n=2}^\infty \frac{1}{n\log n\log\log n} \quad \text{with the integral} \quad \int \frac{dx}{x\log x\log\log x} = \log\log\log x.$$

The series diverges, but so slowly that after 10^{100} terms the sum is still less than 2.

6.11 $\int_1^\infty e^{-x}x^{\alpha-1}\, dx$ is convergent for all α, since $e^x > Kx^{\alpha+1}$ and so $e^{-x}x^{\alpha-1} < 1/Kx^2$ for all positive x. If $\alpha < 1$ then the integral $\int_0^1 e^{-x}x^{\alpha-1}\, dx$ is improper. Since $e^{-x}x^{\alpha-1} \sim x^{\alpha-1}$ as $x \to 0+$, the integral converges if and only if $\alpha > 0$.

To prove the functional equation, integrate by parts:

$$\Gamma(\alpha) = \left[-e^{-x}x^{\alpha-1}\right]_0^\infty - \int_0^\infty -e^{-x}(\alpha-1)x^{\alpha-2}\, dx = 0 + (\alpha-1)\Gamma(\alpha-1).$$

Applying this repeatedly gives $\Gamma(n) = (n-1)!\,\Gamma(1) = (n-1)!$.

6.12 We know that $\lim_{y\to\infty} e^{-ky} y^\alpha = 0$. Substituting $\log x$ for y gives $\lim_{x\to\infty} x^{-k} (\log x)^\alpha = 0$. In this latter limit, substituting $1/t$ for x gives $\lim_{t\to 0+} t^k (\log t)^\alpha = 0$.

6.13 Since $(1/n)\log n \to 0$ as $n \to \infty$, it follows, by applying exp to both sides, that $n^{1/n} \to 1$.

6.14
$$\frac{\log(x \log x)}{\log x} = \frac{\log x + \log\log x}{\log x} = 1 + \frac{\log\log x}{\log x} \to 1$$
as $x \to \infty$.

6.15 By L'Hôpital's rule,
$$\lim_{x\to 0} \frac{a^x - b^x}{c^x - d^x} = \lim_{x\to 0} \frac{a^x \log a - b^x \log b}{c^x \log c - d^x \log d} = \frac{\log a - \log b}{\log c - \log d}.$$

6.16 Substituting $u = st$ in the first integral, we have
$$\int_0^\infty e^{-st} t^n \, dt = \frac{1}{s^{n+1}} \int_0^\infty e^{-u} u^n \, du = \frac{1}{s^{n+1}} \Gamma(n+1) = \frac{n!}{s^{n+1}}$$
by Exercise 6.11.

Call the integrals C and S. Integrating by parts gives
$$C = \left[-\frac{1}{s} e^{-st} \cos at \right]_0^\infty - \int_0^\infty -\frac{1}{s} e^{-st} \cdot - a\sin at \, dt = \frac{1}{s} - \frac{a}{s} S,$$
and
$$S = \left[-\frac{1}{s} e^{-st} \sin at \right]_0^\infty - \int_0^\infty -\frac{1}{s} e^{-st} a \cos at \, dt = \frac{a}{s} C,$$
From the equations $sC = 1 - aS$ and $sS = aC$ the required results follow.

6.17 The result is certainly true for $n = 0$, with $P_0(1/x) = 1$. If we suppose inductively that $f^{(n)}(x) = P_n(1/x) e^{-1/x^2}$, then
$$f^{(n+1)}(x) = e^{-1/x^2} \left(\frac{2}{x^3} P_n(1/x) - \frac{1}{x^2} P_n'(1/x) \right) = P_{n+1}(1/x) e^{-1/x^2},$$
where $P_{n+1}(1/x) = (2/x^3) P_n(1/x) - (1/x^2) P_n'(1/x)$ is a polynomial in $1/x$.

If we now suppose inductively that $f^{(n)}(0) = 0$, then
$$f^{(n+1)}(0) = \lim_{x\to 0} \frac{f^{(n)}(x) - f^{(n)}(0)}{x} = \lim_{x\to 0} \frac{1}{x} P_n(1/x) e^{-1/x^2} = 0.$$

The Taylor–Maclaurin series of f is identically zero, and so cannot possibly converge to f.

6.18 If we put $x = n$ in the inequality $e^x > x^n/n!$ we immediately obtain the desired result.

6.19 Comparing the lower sum, the integral from 1 to n and the upper sum gives

$$\log 1 + \log 2 + \cdots + \log(n-1) \le \Big[x \log x - x\Big]_1^n \le \log 2 + \log 3 + \cdots + \log n.$$

Hence $\log[(n-1)!] \le n \log n - n + 1 \le \log(n!)$, and so $(n-1)! \le n^n/e^{n-1} \le n!$.

Chapter 7

7.1 a) Choose N so that $\|f_n - f\| < \epsilon/2$ and $\|g_n - g\| < \epsilon/2$ for all $n > N$. Then, for all $n > N$, $\|(f_n + g_n) - (f + g)\| \le \|f_n - f\| + \|g_n - g\| < \epsilon$.

  b) Choose N_1 so that $\|f_n\| < \|f\| + 1$ for all $n > N_1$. Choose N_2 so that $\|f_n - f\| < \epsilon/2(\|g\| + 1)$ and $\|g_n - g\| < \epsilon/2(\|f\| + 1)$ for all $n > N_2$. Then, for all $n > \max\{N_1, N_2\}$,

$$\begin{aligned}
\|f_n \cdot g_n - f \cdot g\| &= \|f_n(g_n - g) + (f_n - f)g\| \\
&\le \|f_n\| \, \|g_n - g\| + \|f_n - f\| \, \|g\| \\
&\le (\|f\| + 1)\|g_n - g\| + \|f_n - f\| \, (\|g\| + 1) < \epsilon.
\end{aligned}$$

  c) Since $(f_n) \to f$ uniformly in $[a, b]$, we may choose N_1 so that $|f_n(x)| > \delta/2$ for all $n > N_1$ and all x in $[a, b]$. Hence $|f_n(x)f(x)| > \delta^2/2$ for all $n > N_1$ and all x in $[a, b]$. Choose N_2 so that $\|f_n - f\| < \delta^2\epsilon/2$ for all $n > N_2$. Then, for all $n > \max\{N_1, N_2\}$ and for all x in $[a, b]$,

$$\left| \frac{1}{f_n(x) - f(x)} \right| = \frac{|f_n(x) - f(x)|}{|f_n(x)f(x)|} < \epsilon.$$

7.2 Let

$$f_n(x) = \begin{cases} 1/n & \text{if } x \text{ is rational} \\ 0 & \text{if } x \text{ is irrational.} \end{cases}$$

Then each f_n is discontinuous everywhere, but (f_n) tends uniformly to the zero function as $n \to \infty$.

7.3 It is clear that $f_n \to 0$ pointwise. To show that the convergence is not uniform, we investigate the maximum value of $f_n(x)$ in the interval $[0, 1]$:

$$f_n'(x) = n^3 x^{n-1} - n^2(n+1)x^n = n^2 x^{n-1}\big(n - (n+1)x\big),$$

and so the maximum value, occurring when $x = n/(n+1)$, is

$$n^2 \cdot \left(\frac{n}{n+1}\right)^n \cdot \frac{1}{n+1} = \frac{n^{n+2}}{(n+1)^{n+1}} = n\left(1 - \frac{1}{n+1}\right)^{n+1}. \qquad (9.17)$$

Thus $\|f_n\| \sim n/e$, and so certainly does not tend to zero. Now,

$$\int_0^1 f_n(x)\, dx = n^2 \left[\frac{x^{n+1}}{n+1} - \frac{x^{n+2}}{n+2}\right]_0^1 = \frac{n^2}{(n+1)(n+2)},$$

which tends to 1 as $n \to \infty$. So the integral of the (pointwise) limit is $\int_0^1 0\, dx = 0$, whereas the limit of the integral is 1.

7.4 Again it is clear that $(f_n) \to 0$ pointwise. Here we have $\|f_n\| \to 1/e$, and so the convergence is not uniform. However, $\int_0^1 f_n = n/(n+1)(n+2) \to 0$ as $n \to \infty$.

7.5 Here $\|f_n\| \to 0$ as $n \to \infty$, and so the convergence is uniform. On the other hand, $f_n'(x) = nx^{n-1}(1-x) - x^n$, and (f_n') has pointwise limit g, where

$$g(x) = \begin{cases} 0 & \text{if } 0 \le x < 1 \\ -1 & \text{if } x = 1. \end{cases}$$

Since each f_n' is continuous but g is not, the convergence cannot be uniform.

7.6 a) $f_n(0) = 0$ for all n. If $x > 0$ then $x/(x+n) \to 0$.

  b) Since $f_n'(x) = n/(x+n)^2 > 0$, f_n is increasing in $[0, \infty)$, and so $\|f_n\| = \sup_{[0,b]} |f_n(x)| = b/(b+n)$. This tends to 0 as n tends to ∞, and so the convergence is uniform.

  c) If we redefine $\|f_n\|$ as $\sup_{[0,\infty)} |f_n(x)|$, we see that, for a fixed n, the function $x \mapsto x/(x+n)$ increases steadily towards a limit 1 as $x \to \infty$. So $\|f_n\| = 1$ for all n, and so convergence to 0 is not uniform.

7.7 a) It is clear that $f_n(0) = 0$ for all n, and that, for all $x \ne 0$, $nx/(1+nx) = x/[x + (1/n)] \to 1$.

  b) For all $x \ge b$, $|f_n(x) - 1| = 1/(1 + nx) \le 1/(1 + nb)$. So $\|f_n - 1\| = \sup_{[b,\infty)} |f_n(x) - 1| = 1/(1 + nb) \to 0$ as $n \to \infty$. Thus convergence is uniform in $[b, \infty)$.

  c) $\sup_{[0,\infty)} |f_n(x) - f(x)| \ge |f_n(1/n) - f(1/n)| = 1/2$. So the convergence in $[0, \infty)$ is not uniform.

7.8 $\sup_{\mathbb{R}} |f_n(x) - f(x)| = 1/n \to 0$, and so $(f_n) \to f$ uniformly in $\mathbb{R}$. On the other hand, $\sup_{\mathbb{R}} |(f_n(x))^2 - (f(x))^2| = \sup_{\mathbb{R}} |(2x/n) + (1/n^2)| \ge |(2n/n) + (1/n^2)| \ge 2$, and so the convergence is not uniform.

7.9 For all x in $[0, b]$, $|nx^2/(n^3 + x^3)| \leq na^2/n^3 = a^2/n^2$. Since $\sum_{n=1}^{\infty}(1/n^2)$ is convergent, the given series is uniformly convergent in $[0, b]$ by the Weierstrass test.

7.10 If $x = 1$, then $x^n/(x^n + 1) = 1/2$ for all n, and so the series diverges. If $x > 1$ then $\lim_{n\to\infty}(x^n/(x^n + 1)) = 1$, and so again the series diverges. If $x = 0$ then the series sums trivially to 0. So suppose that $0 < x \leq a < 1$. Then $|x^n/(x^n+1| < x^n \leq a^n$ for all x in $[0, a]$. Since $\sum_{n=1}^{\infty} a^n$ is convergent, the series is uniformly convergent in $[0, a]$ by the Weierstrass M-test.

For all x in $[0, \infty)$, $1/n^2(x+1)^2 \leq 1/n^2$, and so the given series is uniformly convergent in $[0, \infty)$.

$1/(x^n + 1)$ does not tend to zero unless $x > 1$, and so the series is divergent for x in $[0, 1]$. So suppose that $x \geq a > 1$. Then $1/(x^n + 1) \leq 1/(a^n + 1) < 1/a^n$, and so, by the Weierstrass M-test, the series is uniformly convergent in $[a, \infty)$.

7.11 Denote the sum of the series by $S(x)$; then clearly $S(0) = 0$. For $x \neq 0$ we have a convergent geometric series with first term x^2 and common ratio $1/(1 + x^2)$; its sum, after a bit of calculation, is $S(x) = 1 + x^2$. Since S is discontinuous at 0, the convergence in any interval containing 0 cannot be uniform. Consider the set $J = (-\infty, -b] \cup [a, \infty)$, where $a, b > 0$ and where we may assume without essential loss of generality that $a \leq b$. Then, for all x in J, $x^2/(1 + x^2)^n < 1/(1 + x^2)^{n-1} \leq 1/(1 + a^2)^{n-1}$, and so, by the Weierstrass test, the series is uniformly convergent in J.

7.12 For all x in $[0, 1]$, $|x/(n^{3/2} + n^{3/4}x^2)| \leq 1/(n^{3/2} + n^{3/4}x^2) \leq 1/n^{3/2}$. Since $\sum_{n=1}^{\infty} 1/n^{3/2}$ is convergent, the given series is uniformly convergent in $[0, 1]$. This technique will not work for the other series. We obtain the maximum value of $x/(n^{3/4} + n^{3/2}x^2)$ by observing that its derivative, $(n^{3/4} - n^{3/2}x^2)/(n^{3/4} + n^{3/2}x^2)^2$ is zero when $x = n^{-3/8}$. The maximum value is $1/n^{9/8}$. Thus, for all x in $[0, 1]$, $|x/(n^{3/4} + n^{3/2}x^2)| \leq 1/n^{9/8}$, and it follows by the Weierstrass M-test that the series is uniformly convergent in $[0, 1]$.

7.13 For all x in $[0, 1]$, $x^n(1 - x)/n^2 \leq 1/n^2$, and so it is immediate by the Weierstrass M-test that the series is uniformly convergent in $[0, 1]$. This simple argument will not work for the other series. From a simple calculus argument, the maximum value of $x^n(1 - x)$ is obtained when $x = n/(n + 1)$. The value is $[n/(n + 1)]^n[1/(n + 1)] = [1 + (1/n)]^{-n}[1/(n + 1)]$. Now $[1 + (1/n)]^{-n} < 1$ for all $n \geq 1$. Thus, for all $n \geq 1$ and for all x in $[0, 1]$, $x^n(1-x)/n \leq 1/n(n+1)$. Hence the series is uniformly convergent in $[0, 1]$.

7.14 For all x in $[0, 1]$, $\log(n + x) - \log n \leq \log(n + 1) - \log n = \log(1 + (1/n))$.

By the Mean Value Theorem, for all $t > 0$, $\log(1 + t) = \log 1 + t/c = t/c$, where $0 < c < t$. Certainly $\log(1 + (1/n)) < 1/n$, and so, for all x in $[0, 1]$, $(1/n)(\log(n + x) - \log x) \leq 1/n^2$. Hence the series is uniformly convergent in $[0, 1]$.

7.15 (i) $\lim_{n \to \infty}(a_n/a_{n+1}) = 1/2$, so $R = 1/2$. When $x = 1/2$ the series is $\sum_{n=0}^{\infty}(1/(n + 1))$ which diverges; when $x = -1/2$ the series is $\sum_{n=0}^{\infty}(-1)^n(1/(n + 1))$, which converges. So the interval of convergence is $[-1, 1)$. (ii) $\lim_{n \to \infty} |a_n/a_{n+1}| = 1$, so $R = 1$. The series diverges for $x = -1$ and converges for $x = 1$, so the interval of convergence is $(-1, 1]$. (iii) $a_n/a_{n+1} = n!(n + 2)^{n+1}/(n + 1)!(n + 1)^n = [1 + (1/(n + 1))]^{n+1} \to e$ as $n \to \infty$. So $R = e$. When $x = e$, the Stirling formula gives $a_n x^n = n! \, e^n/n^n \asymp n^{n+\frac{1}{2}} e^{-n} e^n/n^n = n^{\frac{1}{2}}$, and so the series diverges at $\pm e$. The interval of convergence is $(-e, e)$. (iv) $a_n^{1/n} = 2 + (1/n) \to 2$ as $n \to \infty$. Hence $R = 2$. When $|x| = 2$, $|a_n x^n| = [2/(2 + (1/n))]^n = [1 + (1/2n)]^{-n} \to e^{-1/2} \neq 0$ as $n \to \infty$. So the interval of convergence is $(-2, 2)$. (v) $a_n/a_{n+1} = (n + 2)\log(n + 2)/(n + 3)\log(n + 3) \to 1$ as $n \to \infty$. So $R = 1$. For $x = 1$ the series diverges (by the integral test); for $x = -1$ it converges by the Leibniz test. So the interval of convergence is $[-1, 1)$. (vi) $a_n/a_{n+1} = (n!)^2(2n + 2)!/(2n)!\,((n + 1)!)^2 = (2n + 1)(2n + 2)/(n + 1)^2 \to 4$ as $n \to \infty$. So $R = 4$. If $|x| = 4$,

$$|a_n x^n| = \frac{(n!)^2 4^n}{(2n)!} \asymp \frac{n^{2n+1} e^{-2n} 2^{2n}}{(2n)^{2n+\frac{1}{2}} e^{-2n}} = n^{1/2} \not\to 0 \text{ as } n \to \infty.$$

Hence the interval of convergence is $(-4, 4)$.

7.16 For $|t| < 1$, $\sum_{n=0}^{\infty} t^n = 1/(1 - t)$. Differentiating term by term gives $\sum_{n=1}^{\infty} nt^{n-1} = 1/(1-t)^2$. Hence $\sum_{n=1}^{\infty} nt^n = t/(1-t)^2$. Finally, integrating term by term gives

$$\sum_{n=1}^{\infty} \frac{n}{n + 1}x^{n+1} = \int_0^x \frac{t\,dt}{(1 - t)^2} = \int_0^x \left(\frac{1}{(1 - t)^2} - \frac{1}{1 - t}\right) dt$$

$$= \left[\frac{1}{1 - t} + \log(1 - t)\right]_0^x = \frac{1}{1 - x} + \log(1 - x) - 1.$$

7.17 $\sin 3x = \sin 2x \cos x + \cos 2x \sin x = 2\sin x(1 - \sin^2 x) + (1 - 2\sin^2 x)\sin x = 3\sin x - 4\sin^3 x$. Hence

$$4\sin^3 x = 3\sin x - \sin 3x = 3\sum_{n=0}^{\infty}(-1)^n \frac{x^{2n+1}}{(2n + 1)!} - \sum_{n=0}^{\infty}(-1)^n \frac{(3x)^{2n+1}}{(2n + 1)!}.$$

It follows that $\sin^3 x = \sum_{n=0}^{\infty} a_{2n+1} x^{2n+1}$, where $a_{2n+1} = (-1)^n(3/4)(1 - 3^{2n})/(2n + 1)!$.

7.18 Since $\sum_{r=0}^{n} x^r$ tends to $1/(1-x)$ as $x \to \infty$ uniformly in any closed interval contained in $(-1,1)$, and since $\sum_{r=1}^{n} a_n x^n$ tends to some function $f(x)$ uniformly in any closed interval contained in $(-R, R)$, it follows that, for all $|x| < \min\{1, R\}$

$$(1 + x + x^2 + \cdots + x^n)(a_0 + a_1 x + a_2 x^2 + \cdots + a_n x^n)$$
$$= (a_0 + (a_0 + a_1)x + (a_0 + a_1 + a_2)x^2 + \cdots + (a_0 + a_1 + \cdots + a_n)x^n$$

tends to $f(x)/(1-x)$ as $n \to \infty$. Thus $\left(1/(1-x)\right) \sum_{n=0}^{\infty} a_n x^n = \sum_{n=0}^{\infty} s_n x^n$. For the last part, note that $\log(1+x) = \sum_{n=0}^{\infty} a_n x^n$, where $a_0 = 0$ and $a_n = (-1)^{n-1}/n$ otherwise.

7.19 $f'(x) = 1/\sqrt{1+x^2}$, $f''(x) = -x/(1+x^2)^{3/2}$, and so $(1+x^2)f''(x) + xf'(x) = 0$ for all x. Differentiating n times by Leibniz's Theorem gives $(1+x^2)f^{(n+2)}(x) + 2nx f^{(n+1)}(x) + n(n-1)f^{(n)}(x) + xf^{(n+1)}(x) + nf^{(n)}(x) = 0$ $(n \geq 0)$, and putting $x = 0$ gives $f^{(n+2)}(0) = -n^2 f^{(n)}(0)$. Since $f(0) = 0$ it follows that $f^{(n)}(0) = 0$ for all even n. Since $f'(0) = 1$, it follows that the coefficient of x^{2n+1} in the Taylor–Maclaurin series is

$$q_n = (-1)^n \frac{(2n-1)^2 \ldots 3^2 . 1^2}{(2n+1)!} = (-1)^n \frac{(2n-1)\ldots 3.1}{(2n+1)2.4\ldots 2n}$$

$$= (-1)^n \frac{(2n)!}{(2n+1)2^2 . 4^2 \ldots (2n)^2} = (-1)^n \frac{(2n)!}{(2n+1)2^{2n}(n!)^2} .$$

Since $q_n/q_{n+1} = (2n+1)^2/(2n+2)(2n+3) \to 1$ as $n \to \infty$, the radius of convergence is 1. Let $|x| = 1$. Then, since

$$q_n \asymp \frac{(2n)^{2n+\frac{1}{2}}e^{-2n}}{(2n+1)2^{2n}n^{2n+1}e^{-2n}} \asymp \frac{1}{n^{3/2}},$$

the interval of convergence is $[-1, 1]$.

Chapter 8

8.1 a) $A(0) = 0$ is clear. For $A(-x) = -A(x)$, substitute $u = -t$ in the integral. A is strictly increasing, since the integrand is positive. It is differentiable in $(-1, 1)$, since the integrand is continuous, and $A'(x) = 1/\sqrt{1-x^2}$ by the fundamental theorem. (This is actually a bit inaccurate, since the integral is improper – which is why I preferred to use the seemimgly more complicated parametric approach.)

b) By Theorems 3.20 and 4.15, $S = A^{-1} : [-\pi/2, \pi/2] \to [-1, 1]$ exists. Since $A(0) = 0$ it follows that $S(0) = 0$. Also, $S'(x) = 1/[A'(A^{-1}(x))] = \sqrt{1 - (A^{-1}(x))^2} = \sqrt{1 - (S(x))^2}$. From this it follows that $S'(x)$ is differentiable in $(-1, 1)$, that $S'(0) = 1$ and that $[S(x)]^2 + [S'(x)]^2 = 1$. Hence

$$S''(x) = (1/2)[1 - (S(x))^2]^{-1/2}(-2S(x)S'(x)) = -S(x).$$

Chapter 9

9.1 Use Stirling's formula: $(1/n)(n!)^{1/n} \sim (1/n)(2\pi n)^{1/2n}(n/e) \to 1/e$.

The Greek Alphabet

α	A	alpha
β	B	beta
γ	Γ	gamma
δ	Δ	delta
ϵ	E	epsilon
ζ	Z	zeta
η	H	eta
θ, ϑ	Θ	theta
ξ	Ξ	xi
ι	I	iota
κ	K	kappa
λ	Λ	lambda
μ	M	mu
ν	N	nu
o	O	omicron
π, ϖ	Π	pi
ρ	P	rho
σ	Σ	sigma
τ	T	tau
υ	Υ	upsilon
ϕ, φ	Φ	phi
ψ	Ψ	psi
χ	X	chi
ω	Ω	omega

Bibliography

[1] P. P. G. Dyke, *An introduction to Laplace transforms and Fourier series.* Springer, 2000.

[2] D. A. R. Wallace, *Groups, rings and fields.* Springer, 1999.

Index

Lightning Source UK Ltd.
Milton Keynes UK
UKHW03f1438070918
328503UK00004B/515/P